SCIENCE & THEOLOGY SINCE COPERNICUS

THE SEARCH FOR UNDERSTANDING

by

Peter Barrett

SCIENCE AND THEOLOGY SINCE COPERNICUS

the search for understanding

by

Peter Barrett

Associate Professor of Physics,
University of Natal, Durban, South Africa

T&T CLARK INTERNATIONAL
A Continuum imprint
LONDON • NEW YORK

T&T CLARK INTERNATIONAL
A Continuum imprint

The Tower Building
11 York Road
London SE1 7NX, UK

15 East 26th St.
New York, NY 10010
USA

www.tandtclark.com

This edition published by T&T Clark International, 2004.

British Library Cataloguing-in-Publication Data
A catalogue record for this book is available from the British Library.

ISBN 0 567 08970 3 (Paperback)
ISBN 0 567 08969 X (Hardback)

Printed and bound in Great Britain by
MPG Books Ltd, Bodmin, Cornwall

PREFACE

This book surveys the rise of modern science and its main developments over the past four centuries; describes some of the associated theological questions and related background thought; outlines briefly the main elements of the present discourse between the natural sciences and Christian theology, and suggests a way to place the scientific world-picture within a wide theistic context of meaning and value.

It was written initially in a much shorter form for a proposed course in the Church History Department of the University of South Africa (Unisa). I hope this expanded version will find general use as an introductory overview of what is a fascinating story of developing ideas and a fast-growing inter-disciplinary subject.

I am especially indebted to John Polkinghorne whose writings have provided my main entry into the Science & Religion discourse. I have valued greatly several conversations with him and with John Brooke, Niels Gregersen, Arthur Peacocke, Michael Poole, Colin Russell and Keith Ward. Neils Gregersen, Colin Russell and Christopher Southgate have offered generously detailed and constructive comments on parts of the text. George Ellis has provided a fine example of the integration of science and ethics, in both thought and praxis, and here I have followed his approach in the forming of an axiomatic scheme that embraces science and theology. John Bowker has long been a distant but influential mentor through breadth of thought and insight and constant reminders that it is issues of temporal power, not theology, that so often control developments in the life of the Church – but that the Science & Religion debate needs to take place nevertheless!

My thanks go to Gerald Pillay (former head of the Church History Department at Unisa) who took the bold step of commissioning this work; also to Cornel Du Toit (Director of Unisa's Research Institute for Theology & Religion) for a congenial partnership in the organizing of the annual conferences of the South African Science & Religion Forum. These have taken place at Unisa since 1993 and have helped to form my ideas.

I write as a retired physicist and a member of the Anglican Church, with a long standing interest in theology and in the missiological and ecumenical aspects of Christianity. Personal background and experience certainly colour one's approach and choice of material, but I hope that at least there is a coherence and unity in this version of a story that needs to be told more widely and developed further. I hope, too, that other scientists will enjoy as much as I have an encounter with some of the historical and philosophical aspects of science that we so often ignore, and that fellow Christians will relax and realise how unthreatening, even enriching, are scientific ideas to the heart of the Christian faith!

Although the field of Science & Religion increasingly involves the human sciences – dealing for example with the mind-brain relationship, the evolution of knowledge and culture, and the wide-ranging question of what it means to be human – no attempt has been made here to include this aspect. Another omission is the story of the Creationism movement in the USA during the 1920s and its wide-spread resurgence since the 1960s, but that topic has of course been treated extensively in the literature.

This work could have been limited to the past two centuries. However, it seemed good to include the Scientific Revolution of the 16th and 17th centuries, not only for its basic importance in the rise of modernity but also to offer South African readers an opportunity to begin to compare the situations in 17th century Europe and 21st century South Africa – each involving the juxtaposition of very different world-views.

Finally, I have taken the liberty of bringing into the last part of the book some of my own thinking about a Trinitarian cosmology – a world-view which is concerned with the search for the transcendental values of truth, goodness and beauty. The search for beauty (often neglected by theologians) forms a key part of the human quest for life in all its fullness and is therefore, perhaps, a place where people can affirm and make common cause with those of other cultures and beliefs.

Peter Barrett
University of Natal, Durban

CONTENTS

1 INTRODUCTION

This book deals briefly with three major shifts in the Western understanding of the world since the mid-16th century, and their significance for the world-view and doctrine of the Christian Church. These shifts are associated with: (i) the Scientific Revolution (16th and 17th centuries), (ii) Darwin's Theory of Evolution (19th century), and (iii) the New Physics (20th century). Each has had profound implications for theology as it seeks to provide a coherent account of the creation, nature and meaning of the universe.

The aim is to give a description of how the work of leading figures in the development of Western science has impacted on Christian belief, from Copernicus to Darwin, then outline some of the significant advances in the physical sciences since Darwin, and finally describe and discuss the lively Science & Religion discourse that has escalated during the past half-century. Altogether it is a story of the search for understanding which, if taken far enough, leads beyond factual knowledge into profound questions of the value and meaning of this richly pluralistic world, and their ethical implications.

As an introduction to the four major sections of the book, we begin with a chronology of leading figures and their work, then describe the main phases of scientific development and theological response, together with the limits of each of these disciplines.

Table 1. Chronology of Leading Figures

Pythagoras	BCE c582 – c507	synthesis of science and mysticism c525
Socrates	469 – 399	wrote nothing, engaged in search for truth
Plato	c427 – 347	*Dialogues* 399 – 347
Aristotle	384 – 322	began 20 years at Plato's Academy 367
Archimedes	c287 – 212	developed classical mechanics and mathematics c240
Augustine	CE 354 – 430	*The City of God* 413-427
Aquinas	1225 – 1274	*Summa Theologiae* 1266 – 1273
Copernicus	1473 – 1543	*De Revolutionibus Orbium Coelestium* 1543
Luther	1483 – 1546	Ninety-five Theses 1517
Calvin	1509 – 1564	*Institutes of the Christian Religion* 1536
Brahe	1546 – 1601	observed supernova 1572
Bacon	1561 – 1626	*Essays* 1597
Galileo	1564 – 1642	*Sidereus Nuncius* 1610 *Dialogue...* 1632; stood trial in 1633
Kepler	1571 – 1630	*Astronomia...* 1609 *Harmonia Mundi* 1619
Descartes	1596 – 1650	*Discourse on Method* 1637
Boyle	1627 – 1691	*New Experiments...* 1660

Ray	1627 – 1705	*The Wisdom of God manifested in the Works of Creation* 1691
Huygens	1629 – 1695	proposed wave theory of light 1678
Newton	1642 – 1727	*Principia Mathematica* 1687
Leibniz	1646 – 1716	*Monadology* 1714
Buffon	1707 – 1788	*Epochs of Nature* 1778
Linnaeus	1707 – 1778	*Systema Naturae* 1735
Hutton	1726 – 1797	*Theory of the Earth* 1795
Paley	1743 – 1805	*Natural Theology* 1802
Lamarck	1744 – 1829	*Zoological Philosophy* 1809
Malthus	1766 – 1834	*Essay on Population* 1798
Cuvier	1769 – 1832	*Preliminary Discourse* 1812
Chalmers	1780 – 1847	*Bridgewater Treatise* 1830s
Buckland	1784 – 1856	*Vindicae Geologicae* 1820
Sedgwick	1785 – 1873	*Discourse on Studies of the University* 1835
Lyell	1797 – 1875	*Principles of Geology* 1830-3
Darwin	1809 – 1882	*On the Origin of Species* 1859 *The Descent of Man* 1871
Huxley	1825 – 1895	essays, critiques, addresses on zoology and on science as new basis of culture
Einstein	1879 – 1955	Special Relativity and the Photoelectric Effect (1905) General Theory of Relativity (1916)
Bohr	1885 – 1962	theory of atomic structure (1913)
Schrödinger	1887 – 1961	wave mechanics (1925)
Heisenberg	1901 – 1976	quantum mechanics (1925), Uncertainty (or Indeterminacy) Principle (1927)

HISTORICAL PHASES OF NATURAL SCIENCE

In the 12th and 13th centuries a strong challenge to the Church's intellectual authority was emerging in the resurgence of Greek natural philosophy and science, especially that of Aristotle.[1] The newly instituted universities in western Europe had acquired Latin translations (from Arabic) of Aristotle's work in physics, cosmology and the nature of motion – to say nothing of his metaphysics, logic and epistemology[2] – and it was the Aristotelian world-picture[3] that was to be superseded in the rise of 'early modern science', commonly known as the Scientific Revolution.

Although as early as the mid-14th century Aristotle's ideas about the nature of motion were under question, the Scientific Revolution is often said to have begun with the publication in 1543 of Copernicus' theory of a sun-centred universe. Thereafter it developed through the formulation of the laws of planetary motion by Kepler (1596); through Galileo's astronomical observations with the newly invented telescope (1609) and his mathematical formulation of accelerated motion; and in the publication of *Principia Mathematica* (1687), the great synthesizing work of Newton on motion and universal gravitation. On the philosophical side, vital stimulation came from the different but complementary ideas of Francis Bacon (in the 1620s) and Descartes (1641). Whereas Bacon advocated the collecting of experimental data and the inductive drawing out of general principles from the data, Descartes looked to the formulation of scientific knowledge by deductive thought from a few fundamental principles.

As a continuation of the bursting forth of new life and vitality of the Renaissance,[4] these scientific developments helped to create that phase of self-confidence and enthusiasm in 18th century European thought known as the Enlightenment, with its high sense of the power of human reason and the bleak picture it drew of an essentially mechanistic universe.

The traditional Christian world-picture was to be challenged again by the rise of the new discipline of geology towards the end of the 18th century, especially the study of rock layers (stratigraphy) and their fossil contents (paleontology). This formed the background to the theory that Charles Darwin built up in the mid-19th century concerning the mechanism by which any biological species can change substantially over long periods. His was a speculative picture and he admitted that it was open to criticism in certain aspects, but it offered a coherent account of a broad range of biological observations, and has since been greatly strengthened through the advent of the science of genetics.

The physics of the 20th century has developed on several fronts, providing the basis of today's rapid advances in technology which have enhanced scientific research and fed a vast field of applications, including those of a military nature. Three thrusts, in particular, have revolutionized our understanding of the physical world: particle physics (the realm of the very small), astrophysics and cosmology (the realm of the very large), and the recently emerging studies of the dynamics of complex systems, including those of a biological nature. This assembly of new insights – which has been called *The New Physics* – has brought a sense of a vast intricately ordered universe, tightly knit at all its levels, and seemingly designed for the development of life.

The new scientific world-picture raises afresh some of the longstanding questions which lie beyond the competence of science – metaphysical questions concerning the existence, creation and destiny of the universe, the nature of divine action, and the meaning of human *being* and *becoming*. These philosophical topics are the concern of *natural theology*.

HISTORICAL PHASES OF NATURAL THEOLOGY

In its philosophical mode Christian theology attempts to understand the nature and significance of the created order, drawing upon other disciplines, especially the sciences, and thence to infer as far as possible the nature of the Creator. This is the task of natural theology.

> If it is true that theology is no mere speculative system but a response to what is, then surely it will always have been in need of

> cool appraisal of the world it seeks to understand. Natural theology – the search for God through the exercise of reason and the inspection of the world – is then not an optional extra, for indulgence by the scientifically inclined, but rather it is an indispensable part of theological inquiry.[5]

At times the term *natural theology* has carried the connotation of a philosophical enterprise that seeks the meaning of the world in terms of simply the universal truths available to human reason and conscience – one that stands against the particularities of a God who acts in history. For the most part, however, natural theology has evolved within the ambit of orthodox Christian belief, complementing the understanding of theology that is based on divine revelation.

Writing in the 13th century, Thomas Aquinas brought together traditional Christian doctrine and carefully selected parts of Aristotelian natural philosophy in his great work, *Summa Theologiae*. Here he built up a cumulative philosophical case (the Five Ways) for the existence of a self-subsistent First Cause, evidence of whose intelligence and purposefulness is to be found in the superb ordering of nature. He appealed to very general facts about the world – such as its existence and its changing nature – and emphasised the exercise of reason as he worked on the two great themes of *existence* (why does something exist rather than nothing?) and *design* (the suggestion of purpose rather than blind chance in the pattern and process of the world).[6]

In the late 17th century a flowering of natural theology began to take place, mainly in England but also in several European countries. It stretched from the writings of Robert Boyle (1627-91) to those of William Paley (1743-1805), and continued well into the 19th century. In contrast to Aquinas, Boyle and others were able to draw upon an array of new empirical knowledge of the cosmos and the world of nature. They could therefore place a much greater emphasis on the evident design of the world, at both the cosmic and biological levels. Newton's *Principia Mathematica* had shown the universality of the force of gravity and inspired a new 'mechanical philosophy' of a cosmos that exhibits mathematical design – a cosmos that Boyle likened to the famous beautifully structured clock in Strasbourg.

This mechanical philosophy raised the problem of how one is to think of the Creator's manner of acting in the world, for a clockwork universe did not seem to leave room for divine manoeuvre. However, the overall picture was considered to preserve the notion of God's sovereignty through the operation of *physical law*, divinely created and sustained. The very concept of physical law, in the sense of a fixed regularity to be observed in natural phenomena, was first introduced into scientific thought in the late 16th century,[7] paving the way for the universe to be seen as a mathematically precise mechanism rather than a spirit-filled organism. In appealing to the clock analogy, Boyle could always invoke God as both the source and sustainer of its motion. However, the mechanical philosophy could readily be viewed in an anti-religious sense, prompting the warning from the French philosopher Blaise Pascal that it was but a short step away from the deism[8] and then atheism that was to emerge in the Enlightenment of the 18th century.

At that time the eminent naturalist John Ray produced one of the classics of natural theology, *The Wisdom of God Manifested in the Works of Creation* (1691), which went to ten editions in forty-four years. This sounded a note of confidence that would persist through the philosophical criticisms of Hume and Kant and the general mood of scepticism of the Enlightenment, reaching a climax in William Paley's *Natural Theology; or Evidences of the Existence and Attributes of the Deity, collected from the Appearances of Nature* (1802). This appeared in nineteen editions in as many years and was one of the most popular works of philosophical theology in the English language.

Following Newton, Paley affirmed the unity of design of the cosmos and its contents as an expression of a single creative mind, but he went beyond Newton, arguing for a God who is beneficent. For example, to the necessity of eating food, God added pleasure. Paley's chapter on the Argument Cumulative opens with the sentence: 'Were there no example in the world, of contrivance, except that of the *eye*, it would be alone sufficient to support the conclusion which we draw from it, as to the necessity of an intelligent Creator'.[9] Likewise, if there were but one watch in the world, it would not be less certain that it had a maker, and every one of a thousand machines would each point that way, with cumulative force. So it is with each superbly functional organ of every plant or animal, allowing Paley to assert confidently: 'It is a happy world after all'.

When Paley's *Natural Theology* made its first appearance in 1802 the discipline of geology was becoming established, especially in England but also in Germany and other parts of Europe. The layered structure of the earth's crust and the fossil contents of rock strata soon indicated, firstly, that the formation of that structure was of a vastly longer duration than the several thousand years calculated from Scripture and, secondly, that the fossil record showed a temporal progression from simpler to more complex creatures. Charles Darwin was involved in geological study and observation, and this picture of earth history formed the background to his slow and cautious development of the theory of evolution.

Despite the incompleteness of the theory, when his famous book *On the Origin of Species by means of Natural Selection* was published in 1859 it cast doubt on the idea of beneficent design in the world of nature, for it provided a plausible alternative explanation – an alternative way of understanding the remarkable well-adaptedness so evident in nature.

Although the book produced no strong polarization of science and religion – indeed there were leading scientists and religious leaders ranged on both sides of the 'evolution' debate – certain major anxieties and criticisms were soon voiced. In the first place Darwin's theory seemed to some fellow scientists to arise from less than rigorous reasoning based on insufficient evidence – there were too many missing pieces in the jigsaw puzzle. Then the questioning of the idea of design in nature created uncertainty about the reality of divine Providence, leading readily to a deistic or even atheistic picture of the world. Darwin's theory could lead, too, to the notion that *homo sapiens* is a species without any special status in the world of nature – a nature 'red in tooth and claw' in which there was no clear place for moral values, and consequently every place for degradation and anarchy. The theory also contradicted the widely held literal interpretation of the creation accounts in the Book of Genesis, thus adding to the already existing questioning of the authority of Scripture.[10]

The long established church denominations have generally combined a respect for the authority of Scripture in the theological and ethical realms with a critical approach to its meaning and interpretation, thereby facilitating a constructive engagement between science and Christian theology. In the early decades of the 20th century, however, there was no particular enthusiasm for such an engagement. Perhaps Karl Barth's firm stand for revelation as the sole key to any understanding of God remained an inhibiting influence for many years.

The progress of 20th century science has brought another flowering of natural theology, especially since the confirmation of the Big Bang model of the universe in 1965. Attention has turned again to the argument from design – that is, the argument for the existence of God – drawn not so much from the ways in which lifeforms are structured but rather from the properties of the universe itself and the laws that underlie them. Indeed, various new perspectives from the realm of the natural and human sciences have raised questions which lie beyond the competence of science and invite responses from the discipline of theology. Consequently, in the Western world there has arisen a fresh wave of discussion over the last three decades on the interaction between theology and science.

The debate has ranged over topics such as: the nature of scientific and religious knowledge; the relationship of the seeming fine-tunedness of the universe (which we discuss later) to the Christian doctrine of creation; the significance of the evolutionary nature of the universe and its life forms; the emergence and nature of the human species, including the nature of the mind/brain relationship; and fresh aspects of the question of how God acts in the world – a world now perceived to be far less mechanistic than previously thought. It is in North America and Europe that much of the activity has taken place, mostly in the area of 'Science & Christian Belief' but steadily broadening into the realm of 'Science & Spirituality' and beginning to attract scholars in other major faiths.

ROLES AND LIMITS OF SCIENCE AND THEOLOGY[11]

In the present phase of natural theology much clarification has come from the setting forth of the areas of competence of science and theology. John Polkinghorne has summarized the differences between the two disciplines as follows:

> (They) are concerned with the exploration of different aspects of human experience: in the one case, our impersonal encounter with a physical world that we transcend; the other, our personal encounter with the One who transcends us. They use different methods: in the one case, the experimental procedure of putting matters to the test; in the other, the commitment of trust which must underlie all personal encounter, whether between ourselves or with the reality of God. They ask different questions: in the one case, how things happen, by what process; in the other, why things happen, to what purpose. Though these are two different questions, yet, the ways we answer them must bear some consonant relationship to each other.

Science and theology have each been accused of arrogance in their claims to know the truth but there is now a growing realism and humility about the limitations of the two disciplines. Science is coming to be seen (at least by many philosophers and sociologists) as a far more relativist project, culturally determined in many of its assumptions and choices of projects, and viewed as just one of the ways in which humans have sought to make sense of their world and manipulate it. Similarly, theologians have come to recognize that their language is much less scientific and much more metaphorical than previously realized. Also, the two disciplines are often seen as complementary ways of seeking understanding of the world, involving mental models of reality which are inevitably incomplete in their representation.

Thus the approach known as *critical realism* (discussed in the section on Epistemic Aspects) has become the working assumption of by far the majority of scientists and theologians – at least those theologians within the community of the Christian faith – rather than the *naive realism* of past ages which claimed exact correspondence between knowledge and the reality to which it refers.[12] Both groups accept that absolutely certain knowledge is simply not attainable since many cultural, personal and conceptual filters intervene between the knowing subject and the object known. At all levels the enterprise of understanding reality has turned out to be open continually to correction and refinement. Unlike the popular stereotype of closed-minded ideologues rigidly defending propositional statements, theologians have as a group been experiencing for some time now a genuine modesty in regard to both what they know and how they know it. Their enterprise of understanding reality has turned out to be as corrigible as that of science.

Nevertheless, scientists and theologians are cautious about aspects of postmodern thought, especially the tendency to embrace varying points of view as equally valid and to deny the existence of an objective reality with which we have to deal.[13] Many see their disciplines as engaged essentially in a quest to know and understand aspects of reality – inferring reality, material or divine, from the way their models and theories make sense of great swathes of physical data on the one hand, or great swathes of spiritual experience on the other,[14] and indeed from the explanatory power displayed by these models and theories. As critical realists they deny that their ideas are no more than constructs of the human mind, without any grounding in reality, not least because of the powerful element of surprise that has accompanied scientific discovery and divine revelation. What human mind could have invented the strange properties of the quantum world, or imagined the scenario of the crucified God?

Science, especially physics, has achieved enormous success as a way of knowing the structure and processes of the material world. It has done so by limiting its realm of inquiry to that which can be empirically investigated, especially through the quest for accurate measurements in appropriately designed experiments on the one hand, and the comparison of such data with predictions from mathematical theories on the other. It has concentrated on physical causes and effects and, as far as possible, avoided speculation about ultimate (divine) causes and purposes.

In contrast, theology has a much wider scope. Its basic task is to seek the deepest level of understanding of whatever is the entire range of reality, taking into account the insights and knowledge derived from other disciplines. Therefore theology is encouraged to be an integrating discipline, drawing upon wide-ranging insights in the formulation of a metaphysical theory of all that is.[15] Some physicists say confidently that a 'theory of everything' will be discovered one day, but it is to theology, not theoretical physics, that one should look for a genuine theory of everything, a meta-narrative which brings together the unfolding world-picture of science and the paradoxes of human experience and understanding. This is a fresh task for natural theology which we shall consider later.

2 THE SCIENTIFIC REVOLUTION: FROM COPERNICUS TO NEWTON

THE MEDIEVAL WORLD-VIEW

The medieval world-picture was based to a large extent upon the natural philosophy of Aristotle – a wide-ranging explanatory scheme which included initially the assumption that the universe is onion-like, comprising concentric crystalline spheres with the earth at the centre. As these rotated about the earth they carried respectively the moon, Mercury, Venus, the sun, Mars, Jupiter, Saturn and the distant stars. The latter lay on the outermost sphere, with relative positions that did not change. This early Greek model, with its simple circular planetary orbits, was clearly inaccurate. It was modified in the mid-2nd century (CE) by Ptolemy who obtained a more realistic description by supposing each of the five planets to move in an epicycle about a point which in turn moved along a circular orbit round the earth. Thus did he account for the fact that the planets appeared periodically to reverse the direction of their motion about the earth and to vary in size.[16]

In Aristotelian thought the cosmos comprised the 'eternal and perfect' realm of the heavenly spheres (the moon and those beyond) and the 'temporal and corrupt' earthly realm (all that lay beneath the moon). Furthermore the notion of *final causes* was all important. If a stone fell to the ground it did so because of its inherent propensity to return to its natural place of rest, and Aristotle would argue that the stone moved more jubilantly every moment because it found itself nearer home.[17] Or if a house was set on fire by lightning, one assumed the wrath of the gods to be the underlying cause. The universe was seen to be essentially organismic, with all processes goal-oriented, arranged to achieve their particular purpose. It was the abandonment of that idea in favour of a mechanistic model that was a central feature of the Scientific Revolution of the seventeenth century. The metaphysical notion of final causes was eliminated in favour of purposeless *efficient* causes, that is, the immediate causes of material processes.

The change did not come easily. It was hard for the human mind to escape from Aristotle's intricate dovetailing of observations and explanations and his appeal to sheer commonsense. What was needed in the first place was a new understanding of the nature of motion. In the Aristotelian world-view nothing moved unless accompanied by a mover.

> (This) was a universe in which unseen hands had to be in constant operation and sublime Intelligences had to roll the planetary spheres around. Alternatively, bodies had to be endowed with souls and aspirations, with a "disposition" to certain kinds of motion, so that matter itself seemed to possess mystical qualities … . The modern theory of motion is the great factor which in the seventeenth century helped to drive the spirits out of the world and opened the way to a universe that ran like a piece of clockwork.[18]

The new understanding of motion – especially that of the earth, moon and planets, and all bodies moving under the action of gravitational force – involved the questioning of both philosophical and theological orthodoxy and it took time to develop, from the publication of Copernicus' work in 1543 to that of Newton in 1687.

At a popular level the prevailing beliefs in Elizabethan England have been summarized thus:

> In the sixteenth century everybody believed in the supernatural; everybody believed, more or less, in magic or the possibilities of magic; everybody believed, to a greater or less extent, in the stars.[19]

Some people sought explanations purely *within* the natural world, drawing upon ideas of astrology and alchemy that not only permeated popular thinking but were fashionable even in intellectual circles.[20] The Renaissance naturalists might have been anxious to reject miracles, but they would still believe that certain plants or stones might bring rain, that animals could prophesy, or that a statue might perspire to announce a grand event.[21] Such aspects of the Renaissance imagination would in due course give way to a different extreme, that is, to the austerity of the mechanistic universe of Newtonian physics. Yet even some of the stalwarts of the emerging mechanical philosophy, including Newton himself, continued privately to be drawn to these older beliefs.

Thus the old organismic and the new mechanistic world-views coexisted and even overlapped over many decades.

THE SCIENTIFIC REVOLUTION AND ITS BACKGROUND

It is widely accepted that 'modern science' arose in the Europe of the 17th century (towards the end of the Renaissance), introducing a new understanding of the natural world. 'Is it not evident in these last hundred years ... that almost a new Nature has been reveal'd to us?' wrote the poet John Dryden in 1668, even before the arrival of Newtonian cosmology. What was this revolutionary modern science and what were the underlying factors, both within and beyond Europe, that encouraged its emergence *then* and *there*? These are questions that have led to a large body of scholarship since the early years of the 20th century.

It has been suggested that the word *science* began to displace the term *natural philosophy* in the late 18th century, as the study of nature acquired a more technical, less humanly inclined ethos. However, in a recent major historiographical review[22] the Dutch science historian Floris Cohen describes the 17th century Scientific Revolution itself as essentially the transition from natural philosophy to science – one in which metaphysical explanations of the natural world (such as Aristotle's) were set aside in favour of a much more empirical piecemeal approach. Galileo, for example, developed a theory of projectile motion without reference to any contemporary world-view. The acquisition of precise quantitative understanding of a limited aspect of nature became a hallmark of the new science – and precision meant the measurement of shape, size and quantity in order to formulate mathematical descriptions of natural phenomena, especially the phenomenon of motion.

Serious questioning of Aristotelian physics can be traced back to the 14th century and even to some early Greek philosophers, but it can be argued that modern science began to take shape in the work of Kepler and Galileo around the year 1600, the first two scientists (then called *natural philosophers*) truly to mathematize nature.[23]

The central features of the new science have been listed as: the mechanistic model of nature (referred to above); the emphasis on unprejudiced observation; and the deliberate setting up of experiments designed to focus on a limited aspect of nature and seek answers to precise questions,[24] seeking where possible to formulate mathematical descriptions of nature's phenomena, without reference to notions of first causes or ultimate purposes. Science historian Alexandre Koyre regarded the birth of modern science as essentially the transition from the world of the 'more-or-less' to the universe of precision.[25]

For Koyre the transition signalled too a change in world-view, of which modern physical science was at once the expression and the fruit. Fundamental propositions such as the principle of inertia and Newton's second law of motion took their place in the framework of a larger transition, a fundamentally new overall conception of *motion*. And this transition, in its turn, could only take place in the even wider framework of a new conception of the universe itself. He referred to this transition as an 'intellectual mutation', the change from movement as a goal-oriented process to movement as a value-neutral state of a body.[26]

Cambridge historian Herbert Butterfield estimated the significance of this intellectual mutation in the oft-quoted remark:

> Since that revolution overturned the authority in science not only of the middle ages but of the ancient world – since it ended not only in the eclipse of scholastic philosophy but in the destruction of Aristotelian physics – it outshines everything since the rise of Christianity and reduces the Renaissance and Reformation to the rank of mere episodes, mere internal displacements, within the system of medieval Christendom.[27]

This may seem an overstatement[28] but it created a sense of joyous recognition in the mind of Floris Cohen and is echoed in the following remarks of science historian Richard Westfall:

> The Scientific Revolution was the most important 'event' in Western history .... For good and ill, science stands at the center of every dimension of modern life. It has shaped most of the categories in terms of which we think, and in the process has frequently subverted humanistic concepts that furnished the sinews of our civilization. Through its influence on technology it has helped to lift the burden of poverty from much of the Western world, but in doing so has accelerated our exploitation of the world's finite resources ... . Through its transformation of medicine, science has removed the constant presence of illness and pain, but it has also produced toxic materials that poison the environment and weapons that threaten us with extinction .... I am convinced that the list describes a large part of the reality of the late twentieth century and that nothing on it is thinkable without the Scientific Revolution of the sixteenth and seventeenth centuries.[29]

What then were the main causes and circumstances of this landmark in human history if it is to be regarded as not simply a spontaneous 'intellectual mutation'? Or, to put it in the negative form of the question pursued by the renowned sinologist[30] Joseph Needham during the period 1954-1981: What, if anything, can we learn about the causes of the Scientific Revolution by considering the fact that modern science did not emerge in any of the other great civilizations of the past, especially those of China, India and the Islamic world?[31] This is a complex and wide ranging question – one of the great questions facing historians, claimed Needham – and it begs answers in terms of attitudes towards the natural world on the one hand and socio-economic factors on the other. It has been stimulated by the multi-faceted comparative approach that Needham brought to it as he immersed himself sympathetically in a culture vastly different from his own, for which he became an ardent advocate.

But first, how strongly developed was pre-modern science in these three cultures, and what contribution, if any, did they make to the Scientific Revolution? We consider each of them briefly, giving particular attention to the place of early Islamic science before returning to the situation within Europe itself.

Needham maintained that China gave much more to the world than it received, both in technical invention and in scientific discovery. He tried hard to press that claim in the areas of magnetism, medicine, cosmology and observational astronomy, and in the measurement of time. However, his exhaustive searches seem to have uncovered little exchange of *scientific* knowledge, rather than expertise of a purely technical nature, and in the end he admitted that Western science developed on the whole without any obvious contribution from China.[32] He regarded this as a problem of the communication of abstract ideas:

> The mutual incomprehensibility of the ethnically-bound concept systems did severely restrict possible contacts and transmissions in the realm of scientific ideas. This is why technological elements spread widely through the length and breadth of the Old World, while scientific elements for the most part failed to do so.[33]

In the case of India excellent work was carried out in mathematics and astronomy but there is little evidence of activity in other physical sciences (as distinguished from botany, medicine and technology). Again there was a problem of non-transmission to the West. India's main scientific work was no doubt well known within the realm of early Islamic science, which was itself enriched by Indian mathematics and astronomy. However, the overall contribution from India was not included in the succeeding translation of scientific works from Arabic into Latin. For example the grand survey of Indian science by the Persian astronomer/ physicist al-Biruni, written round about 1012, was not translated into a European language until 1888.[35]

This leaves early Islamic science (sometimes referred to as Arabic science) as the main external contributor to the rise of science in Europe.

In the early centuries of Islamic society scientific research and education developed through four types of institution not previously found in Greek or Roman civilization: the hospital, the public library, the *madrasa* (a school of higher learning, especially religious learning), and the astronomical observatory. This enterprise started as a vast and vigorous enterprise of translating Greek science and philosophy into Arabic, from the mid-8th century to the early 10th century – what the historian of Islamic science A.I. Sabra described as a major feat of the imagination, of organization, and of scholarly excellence.[36] This led to advances in geometry and algebra, in astronomy, in methods of determining material densities and alloy compositions, and in optics – to say nothing of an intensive development of alchemy (the forerunner of chemistry) and medicine. The work in optics included for example the geometry of vision and, at a later stage, the mathematical analysis of the rainbow (with the aid of water-filled glass spheres as models of raindrops).[37]

After that period of high growth – especially the so-called Golden Age of the 9th and 10th centuries with its cluster of scientific luminaries – scientific development continued in a lower key, even as the same ancient learning was being re-translated from Arabic into Latin and given an orderliness and unity that it had lacked previously.[38] Here lay Islam's main contribution to the Scientific Revolution, and Arabic words such as algebra, algorithm, zenith, azimuth, alcohol and alkali are a reminder of the crucial role it played.

But this innovative scientific enterprise, expressive of a rich and largely autonomous civilization and seemingly so full of promise, began to stagnate and even to decline from about the end of the 14th century – or much earlier in the view of Turkish science historian Aydin Sayili. How and why did this happen, in such contrast to the development of science in Europe? This was the fruitful question addressed by Sayili in a dispassionate analysis of that decline.[39] In the writings of Sayili and two American scholars, G.E. von Grunebaum and J.J. Saunders, we are confronted with varying speculative interpretations in terms of religious and intellectual positions on the one hand and social, political and economic factors on the other. We also note a view from within Islamic orthodoxy, that of Iranian science historian Seyyed Hossein Nasr who is also a Sufi philosopher with an American degree in physics.

Since the impact of new knowledge on religious belief is a recurring phenomenon, it seems worth noting the picture presented by these scholars.[40] There are surely parallels to be found and differences to be noted, not only in the rise of science in Europe but wherever modern science confronts traditional world-views.

In the early years of Islamic science the new knowledge that became available from its Greek heritage was quite well received, valued for its medicine and astrology and for its contribution to such tasks as navigation across sea and desert and the setting up of the calendar. Greek philosophy too was appreciated for its help in placing the data of religion in an appropriate body of rationalized thought. However as major differences between Islamic faith and Greek thought became clear, many theologians gradually began to frown upon the newly acquired knowledge. This body of philosophy was increasingly regarded as alien to Islam; so too was the science linked to it. Thus there developed a dichotomy between the ancient Greek sciences and the new Islamic science – but even the latter was valued only insofar as it served the community of the faithful.

Islam did not readily make room for straightforward curiosity about nature, as did classical Greece, nor for the early Western idea of science as a means to gain insight into the handiwork of creation and thereby glorify God. Indeed it has been suggested that within the practice of early Islamic science there remained a nagging sense of the impropriety of delving into the mysterious ways of that divine handiwork. G E von Grunebaum offers the following interpretation:

> (T)he sciences never did shed the suspicion of bordering on the impious which, to the strict, would be near-identical with the religiously uncalled-for. This is why the pursuit of the natural sciences, as that of philosophy, tended to become located in relatively small and esoteric circles .... It is not so much the constant struggle which their representatives found themselves involved in against the apprehensive skepticism of the orthodox which in the end smothered the progress of their work – rather it was the fact ... that their researches had nothing to give to their community which this community could accept as an essential enrichment of their lives[41]

an assessment largely shared by Sayili. No Scientific Revolution is conceivable, adds Floris Cohen, unless an at least semiprofessional cluster of scientists exude a deep and abiding belief in the value and dignity of their work.[42]

Among socio-political factors the foremost was undoubtedly the impact of the four waves of destruction by the Mongols that the Islamic world suffered from the 11th century onwards, just as Western Europe was emerging from the Dark Ages and beginning to build up a new civilization.[43] For centuries there had been a giant area of free commerce controlled by the Arabs, stretching from Senegal to Canton, and then the invasions broke it up, changed the social structure and brought economic decline. Saunders summarized the situation thus:

> The contrast with the West could hardly have been greater. In the Muslim world there was no hereditary aristocracy, no clergy or monks, and no Third Estate (middle class). As commerce declined, the merchant level sank to the level of the artisans and craftsmen, and society resumed its old simple structure of landlords, officials and peasants.[44]

The result was that a free, tolerant, inquiring and open society gave place to a more rigid and closed society in which the progress of secular knowledge was slowly stifled. It is not surprising if there was now a closing of the ranks and a turning inwards to find consolation in religion.

However, a further element in that scene is suggested by Seyyed Hossein Nasr who has argued that toward the end of the Middle Ages, Islamic civilization declined to take the necessary steps toward a building up of science because of the feeling that no good could come in the end from *scientia* (knowledge) devoid of *sapientia* (wisdom). Affirming the fierce attack against science by the renowned 10th century Islamic theologian al-Ghazzali, Muslim scholars recoiled from the vision of an autonomous triumphant science lacking the constraints of wisdom. Whether or not this was a conscious decision, Nasr's interpretation points to a distinction between the respective world-views of Islam and post-medieval Christianity. He wrote:

> The main reason why modern science never arose in China or Islam is precisely because of the presence of metaphysical doctrine and a traditional religious structure which refused to make a profane thing of nature... . The most basic reason (why the scientific revolution did not develop elsewhere) is that neither in Islam, nor India nor the Far East was the substance and stuff of nature so depleted of a sacramental and spiritual character, nor was the intellectual dimension of these traditions so enfeebled as to enable a purely secular science of nature and a secular philosophy to develop outside the matrix of the traditional intellectual orthodoxy.[45]

Elsewhere he writes of Islamic science as a continuing project, drawing from earlier scientific traditions but seeking always to be rooted in the Quran and especially in its central doctrine of *tawhīd* – 'being one' or 'making one'. The aim is to show the unity and interrelatedness of all that exists, the unity of nature being an image of the unity of the Divine Principle.[46] Furthermore, within such a framework causality is not confined to the physical realm, for there is also vertical causality emanating from the will of God, a point which links with ideas discussed later in the section on Divine Action.

We return now to the situation in Renaissance Europe where the world-view of traditional Christian faith was perhaps less sharply drawn and more flexible in its response to the new understandings of the world.

Much scholarship has been devoted to the groundwork provided by Greek, medieval and Renaissance thought about the natural world; also to the effects of Puritanism and the Reformation; to the arts and crafts and their developing technologies; to the rise of capitalism, the Voyages of Discovery and the printing press.[47]

In such analysis it has been claimed by some scholars that the climate of thought within English Puritanism was a key factor in the rise of early modern science, at least as it arose in England. This notion – known to historians of science as 'Merton's thesis'[48] – is regarded by other scholars as an undue simplification. It is criticized for its overly broad definition of Puritanism, its description of science in highly empirical utilitarian terms (far removed from the theoretical contributions of Kepler, Galileo and Newton), and its underplaying of the scientific work in contemporary Catholic cultures.[49]

Reijer Hooykaas argues, with Merton, that Protestantism's high regard for manual labour – for the work of intellectual artisans such as engineers and architects – provided the means and climate of social acceptability that encouraged experimental science. Joseph Ben-David suggests, similarly, that the social conditions of northern Europe were such as to legitimize the work of science and thus to encourage the doing of science as a vocation in its own right.[50] That ethos was absent in post-Socratic Greece, for example, where manual work other than that of a military or agricultural nature tended to be left to slaves or free artisans.[51]

No doubt the social ethos in northern Europe exerted a major influence on the development of science, but what of the ways of thought needed for such a change? – a question one might ask in response to Needham's remark about the lack of any impetus to Chinese science from the direction of Buddhist thought:

> One of the preconditions absolutely necessary for the development of science is an acceptance of Nature, not a turning away from her. If the scientist passes the beauty (of Nature) by, it is only because he is entranced by the mechanism. But other-worldly rejection of this world seems to be formally and psychologically incompatible with the development of science.[52]

John Bowker makes the same point when he contrasts the 'inversive' systems of human thought which look for profound experience and truth *within* the human mind – Buddhism for example – and the 'extraversive' systems which value the human body's relatedness to the outside world (as well as attending to the interior life) – Christianity or Islam for example. He writes that

> for a culture with an emphasis on inversive exploration, the extraversive exploration of environments is not a valued priority; and it is unsurprising that the natural sciences emerged from specifically Muslim and Christian roots.[53]

It has been asserted too by Christian apologist and missiologist Lesslie Newbigin that the understanding of the created order as both *rational* and *contingent* (not *necessarily* the way it is but deriving from the unfettered choice of the Creator) – therefore discoverable not by philosophical endeavour but by direct theory-assisted experimental investigation – constituted a philosophical prequisite for the scientific Revolution that was especially characteristic of the Judaeo-Christian world-view.[54]

Altogether there seems to be no single clearcut underlying cause of this new phase in human consciousness and understanding. Indeed, given its sheer brilliance and complexity, one should perhaps expect there to be a large element of contingency in the rise of modern science. The historian of technology Lynn White expresses it thus:

> A complex society like that of traditional China or that of medieval Europe .... is composed of a vast web of activities, ideas and emotions. Innovation can start anywhere in this network and then travel in the most unexpected ways and directions, producing results that could scarcely have been guessed before the fact.[55]

Here we shall follow in outline the main features of this unprecedented phenomenon through the contributions of some of its key figures, noting certain theological issues with which they were concerned. Copernicus, Kepler and Galileo helped to establish the heliocentric lay-out of the universe; Bacon led the way in establishing the new science as an empirically based enterprise; Descartes gave it a new metaphysical basis and, with Boyle, encouraged a view of the natural world as a mechanism;

and Newton provided the framework in which to some extent these different contributions could be unified.

NICOLAUS COPERNICUS (1473-1543)

Although there were earlier contributors to the Scientific Revolution, some writers see Copernicus' revision of Ptolemy's celestial system as the decisive initial step in which the earth was no longer regarded as the stationary centre of the cosmos but simply another planet, orbiting the sun. This required a huge change in perception of the universe.[56] However, the physical aspects of his world-view remained entangled with concepts of value, teleological explanations and forms of what we should call animism – the attribution of a living soul or life-force to plants, inanimate objects and natural phenomena – and so Copernicus can be regarded as closing an old epoch much more clearly than opening a new one.[58] Like many throughout the 16th and 17th centuries, he still shared much of the ways of thinking of the Renaissance. Consequently his hypothesis may be regarded more as a significant *preliminary* step which prepared the way for the work of Kepler and Galileo around the year 1600.

Nicolaus Copernicus received a good education at the University of Cracow (Poland), especially in mathematics and astronomy. Then, to improve his career prospects, he went to Italy for training in medicine and canon law and acquired a doctorate in the latter. Thereafter he became a lay canon of a cathedral in Prussia in what was a senior administrative post. At the same time he did a modest amount of observational astronomy and looked to improve the way the known data were used to calculate planetary orbits. There were a number of deficiencies and a general untidiness in the ancient Ptolemaic system that must have prompted his search for a simpler and more accurate model. At that time, too, the position of astronomy was being questioned: Was the study of heavenly motions simply a utilitarian enterprise for the purposes of astrology and the fixing of festival dates, to be carried out in ever-increasing detail by 'mathematicians' – men who performed the calculations for astronomy – or was it also the concern of the 'philosophers', whose task was to inquire into the true nature or *physis* of things, the way the physical universe actually is? Traditionally it had been the latter who held the higher intellectual status but, whatever his concern in this respect, he seems to have believed that his model was more than a calculating device – that it represented physical reality.

Possibly with the encouragement of ancient Greek speculation about a heliocentric universe and a neoplatonist desire for perfect mathematical forms,[59] Copernicus used not only physical but also metaphysical arguments for his cosmological model. He referred to the beauty, harmony and divine order of the cosmos. The sun is given the place of rest, and hence honour:

> In the centre of all rests the sun. For who would place this lamp of a very beautiful temple in another or better place than this from which it can illuminate everything at the same time? . . . . And so the Sun, as if resting on a kingly throne, governs the family of stars which wheel around.[60]

For Plato and Aristotle the perfection of the circle meant that all celestial orbits must be circles – or combinations of circles – and Copernicus remained so blinkered by that presupposition that he introduced at least as many epicycles as Ptolemy to match the astronomical data. Butterfield referred to his scheme as 'that fantasia of circles and spheres'. In due course it was shorn of some of its complications and came to be recognized as a huge achievement, with a certain unity and simplicity that appealed instantly to those prepared to make the considerable mental leap necessary.

Copernicus' system of purely circular uniform motions may have been mathematically simpler than Ptolemy's but it was incompatible with Aristotelian ideas and seemed absurd to many of his fellow astronomers. Although he saw it as a restoration of ancient astronomy he could yet write: 'The scorn which I had to fear on account of the novelty and absurdity of my opinion almost drove me to abandon a work already undertaken'[61] and, like Darwin three centuries later, he was long reluctant to publish his work. The key idea was that the earth rotates once daily on its axis and revolves once annually round the sun, contradicting Aristotelian natural philosophy, which claimed that only heavenly bodies could have constant circular motion. It seemed a matter of common sense, too, that if the earth rotated clouds would float westwards and freely falling objects would not fall vertically. Furthermore, if the earth were not at the cosmic centre – that is, at the natural resting place to which all material bodies were innately drawn – loose objects would fall away from the earth's surface. As yet there was little conception of gravitational attraction.

Apart from such questions of physical behaviour the Copernican theory came to be seen as a threat to the whole Christian world-view. It stood against Aristotle in claiming that the earth is simply another planet and seemed thus to demote man from the central position in the created order. The latter notion has indeed been read into Copernicus' world-view, even in recent years.[62] However Copernicus emphasised that the earth, and even its orbit, is a mere point in this universe that is so vast that no astronomer had yet observed any stellar parallax (that is, any change in the observed relative positions of the stars as the earth moves). Whether the earth is at the centre or at one earth-orbit-radius away from the centre was neither here nor there. Furthermore he would no doubt have argued that the primacy of man rested not so much on the centrality of his location as on his being made in God's image and on the affirmation of God's love in the Incarnation of Christ ('the Word made flesh'). For Copernicus a heliocentric world-picture was clearly compatible with an anthropocentric world-view.

As a loyal member of the Catholic Church, Copernicus hesitated for a long time to publish his work, thinking not only of the scorn it might evoke in fellow astronomers but also its potential for theological controversy. He did, however, allow his keen disciple, Joachim Rheticus (a Lutheran mathematics professor from the University of Wittenberg), to publish his own account of these ideas in 1540. It then became easier, no doubt, for Copernicus to consent to the publication of the entire six-volume work, *On the Revolutions of the Heavenly Spheres*, which he dedicated to the Pope, Paul III, expressing the hope that it would receive the Pope's understanding and approval. For the author the new scheme was not just a mathematical tool for calculating celestial motions but described the actual lay-out of the cosmos, and at this late stage of his life he made no attempt to hide his unorthodoxy: 'What I have accomplished in this matter I leave to the judgement of Your Holiness in particular and to that of all other learned mathematicians'.[63]

Publication did not get under way until immediately after the author's death in 1543. It has been estimated that about five hundred copies of that first edition were made – followed by further editions in 1566 and 1617 – and these were distributed through Europe during the next few decades, in both Catholic and Protestant areas.[64] Yet there was no immediate backlash. If the author himself believed the physical reality of his model, others were free to treat it as no more than an astronomical tool. It was more than seventy years before the storm broke around Galileo.

In the mean time the Copernican model found its way slowly into the thinking of many individuals. Several new books supported it. If there were few objections initially, that may mean that these were so self-evident that the writers (often poets or philosophers) declined to state the obvious. The respected Aristotelian Professor of Greek at the University of Wittenberg, Philip Melanchthon, opposed the Copernican teaching brought by Rheticus as anti-Aristotelian and contrary to biblical passages in which it is the sun, not the earth, that moves. But he was not unduly critical of Copernicus and could accept him as a moderate reformer for, as a Lutheran, Melanchthon himself was a rebel against the authority of Rome. On the other hand the Swiss theologian of the Reformation, John Calvin (1509-1564), did not interpret such passages literally. He emphasised the authority of Scripture but, like St Augustine, accepted the idea that in Scripture the Holy Spirit 'accommodated himself' to the understanding of the common man. If the sun is said to move, that statement is merely a reflection of the common perception. 'Moses wrote in a popular way', remarked Calvin. 'One should not look there for astronomy and other abstruse sciences; it is a book for laymen'.[65] Calvin encouraged the study of science, regarding astronomy as an art that 'unfolds the admirable wisdom of God'.

Copernicanism remained free to spread throughout Europe for several decades and it was not till 1616 that *De Revolutionibus* began to receive the serious attention of the Church's censors.

The 1570s were important years for astronomy and, ultimately, for the confirmation of the Copernican system. Firstly there was the great supernova that flared up in 1572 – an explosion that remained visible to the naked eye for two years and, from the smallness of its parallax, clearly lay far beyond the moon where all was supposed to be eternally flawless and unchanging. Another supralunar phenomenon was the spectacular comet that appeared for two months in 1577, seeming to travel through the crystalline spheres of the old celestial model. This, too, contradicted the notion of the changelessness of the heavens. It was during this time that the Danish astronomer, Tycho Brahe (1546-1601), was steadily building up a collection of precise astronomical data – the best available until the telescope was invented about the year 1609. Near the end of the century Brahe retired to Prague where he received the young mathematically gifted astronomer Johannes Kepler as his pupil and passed on to him this wealth of unarranged material.

JOHANNES KEPLER (1571-1630)

Working over many years with an almost mystical fervour, Kepler brought order into Brahe's data, searching always for underlying mathematical relationships. Drawing upon his mathematical expertise in the subject of conic sections, he saw eventually the patterns which he was able to express in the form of his famous three laws of planetary motion.

Kepler's childhood was marked by poverty, unsettled family life and frequent illness. At the age of thirteen he won a scholarship to a Lutheran seminary which brought him a stable and rigorously disciplined routine, but also bullying by fellow students who regarded him as an intolerable egghead and beat him up at every opportunity.[66] He was a religious young man but also quarrelsome and regarded by his teachers as theologically suspect. Subsequently a post as a teacher of mathematics and astronomy in Austria opened up a career that was clearly more appropriate for him than the role of a humble clergyman that he had previously envisaged. He was to become the first great Protestant scientist.

When Kepler learned about Copernicus' heliocentric system of the cosmos, he accepted it immediately. In the ensuing years he brought to his astronomy a lively imagination and the platonist assumption that the world is patterned on perfect forms. He looked especially for beauty in the geometry of the universe and was no doubt delighted to find that the relative spacings of the planet-carrying crystalline spheres were such that the five known equal-faced geometrical shapes – the cube (six faces), pyramid (four), dodecahedron (twelve), icosahedron (twenty), and octahedron (eight) – could be chosen in such an order that they fitted tightly between the crystalline spheres. Here indeed was a sign of a master-mind. 'God, like a master-builder, has laid the foundation of the world according to law and order' wrote Kepler. 'Our piety is the deeper', he added, 'the greater is our awareness of creation and its grandeur'.[67] However, the pattern of the cosmos did not yield easily to his geometrical imaginings.

After struggling on and off for several years to fit Brahe's data to various combinations of circular orbits, Kepler was compelled to abandon the centuries-old obsession with *circular* motion. His respect for the facticity of nature meant that a discrepancy of a mere eight minutes of arc between the observed and calculated positions of the planet Mars was sufficient to require him to produce a new model of planetary orbits. 'I was almost driven to madness in considering and calculating this matter', he wrote, but in the end he discovered that the orbit of Mars was an ellipse. After centuries of belief in the circle as a symbol of heavenly perfection he made this discovery with great reluctance. 'I have laid a monstrous egg', he remarked.

Kepler's discovery can be viewed as a triumph of empiricism over dogmatism. Nevertheless it reflected the deep connection between scientific and theological thinking that was characteristic of the age. In that the ellipse could be said to be a mixture of a circle (symbolising the spiritual) and a straight line (associated with the material), Kepler regarded an elliptical orbit as one in which the ideal circularity has been diluted with a small degree of straightness, a consequence of the materiality of the planets. Furthermore, he insisted that planetary motions were the result of a real *physical* force, emanating from the sun, thus providing the germ of an idea for Newton's work on gravitation.

For Kepler geometry provided a powerful key to understanding the entire created world. Indeed, he claimed that 'geometry existed before the

Creation, is co-eternal with the mind of God, *is God himself* . . . . geometry provided God with a model for the Creation'. 'It is absolutely necessary,' he continued, 'that the work of such a perfect Creator should be of the greatest beauty'. The idea of cosmic beauty and harmony continually underpinned his search for mathematical relationships in the planetary data.[68]

Ultimately he discovered three fundamental laws of planetary motion:

- Each planet moves in an elliptical orbit with the sun at a focus of the ellipse.
- The line between the sun and any planet marks out equal areas in equal times.
- The square of the time for a planet to travel one orbit is proportional to the cube of its mean distance from the sun.

Another part of Kepler's theological concern was that Scripture be interpreted aright, so that no conflict should appear between the 'book of Scripture' and the 'book of Nature'. If the Reformation saw Protestants and Catholics striving to outdo each other in defending the authority of Scripture, this did not inhibit criticism of naive literalism. Indeed, writing in the same vein as Calvin, Kepler claimed that since Scripture is addressed to both 'scientific and ignorant men', it is bound to be worded in accordance with human senses, as when the sun seems to move across the sky (see Psalm 19 and Joshua 10 for example). Again, 'if anyone alleges on the basis of Psalm 24 . . . that the earth is floating on the waters, may it not be rightly said to him that he ought to set free the Holy Spirit and should not drag Him into the schools of physics to make a fool of Him?'.[69]

Perhaps it was Kepler's respect for the biblical picture of a created, therefore contingent, universe that eventually drew him away from his cherished notion of magic numbers in nature and enabled him to begin to see mathematical relationships instead as the key to nature's processes. In this respect he was the crucial link between Copernicus on the one hand and Galileo, Descartes and Newton on the other. The deductive mathematical approach of these men was complemented by the empirical experimental approach associated with figures such as Bacon, Gilbert and Boyle. These two streams in the development of the Scientific Revolution converged in the brilliant work of Newton although, ultimately, it was the empirical that constituted the main criterion in Newton's search to understand the physical world.[70]

GALILEO GALILEI (1564-1642)

Galileo was Kepler's immediate successor in the task of applying mathematics to the processes of nature. Despite his early promise as a student at the University of Pisa, he was not granted the scholarship he needed there. He worked at home for a while, developing a flair for making new mechanical devices. In due course, through the patronage of Cardinal Del Monte, he obtained a mathematics lectureship at his old university and thence a professorship at the University of Padua, a stronghold of Aristotelian philosophy. It was here that he did much of his work on accelerated motion, questioning Aristotle's assertion that bodies fall to earth with velocities proportional to their sizes. This was the work that was perhaps his main legacy to modern science, which he wrote up during the last few years of his life while under house arrest.

While at Padua, Galileo embraced Copernicus' heliocentric model of the cosmos, believing that evidence for the earth's motion lay in the existence of the tides – the oceans sloshing back and forth like water in a bucket when it is transported. There was as yet no concept of the force of gravity acting across empty space such as, for example, the pull of the moon on the earth's oceans. But, as he explained to Kepler (in the year 1596), he had refrained from publishing his Copernican ideas because of the 'ridicule and derision Copernicus our teacher' had received. He was concerned about strong opposition from university colleagues who had staked their scholarship on Aristotelian philosophy.

By 1610 the situation was different. In quick succession Galileo had made one of the first telescopes and discovered astronomical features that contradicted both Aristotelian physics and the Ptolemaic system of the cosmos. Thereafter he promoted the Copernican system vigorously, but not entirely rigorously. When Galileo looked at the heavens with his telescope he was amazed. With a magnification of thirty, he saw a lunar surface of mountains and valleys – not the unblemished globe that had been imagined for so long as part of the perfection of the heavens. The stars of the Milky Way were seen in much vaster numbers, and then came the surprise of Jupiter's four moons, suggesting that the moon-carrying earth, too, could be simply another planet. Within a few more months, he observed the planet Venus going through changes of phase such as those of the moon – from a dark disc, through a crescent shape, to a fully illuminated disc – in contradiction of the Aristotelian assumption of celestial changelessness.

Galileo unjustifiably made the claim that the phases of Venus *proved* that the sun is the stationary centre of the cosmos. As Kepler pointed out, heliocentricity would imply Venus' behaviour, certainly, but the latter does not necessarily imply the former. Brahe, for example, had assumed the identical *relative* motions of earth, planets and sun as in the Copernican model, but with the earth stationary instead of the sun, and this, too, would lead to the changing phases of Venus. Also, Brahe's scheme accorded with common sense and with the plain wording of Scripture.

Galileo ventured to explain to his fellow Catholics, clerical and lay, that the heliocentric system was consistent with a reasonable interpretation of Scripture. For a time he was encouraged in this stance by senior ecclesiastics in Rome and elsewhere, but the idea of a moving earth was a problem for theology, tied as it was to Aristotle. If the Aristotelian-based world-view were to unravel, would this not take the entire contemporary Christian teaching with it? Besides, here was a lay professor venturing into the preserves of the theologians, however reasonable and important the issues he raised.

In 1615 Galileo presented his ideas on the relationship between science and theology in a long letter to the Grand Duchess Christina. As he intended, the contents were circulated and eventually reached Cardinal Bellarmine, the senior theologian and guardian of orthodoxy at the Vatican. Galileo's concern was to persuade the Church of the folly of its anti-Copernican stance.

He was conscious of a long tradition of well meaning defence of the authority of Scripture. He quoted Copernicus' critical description of the 6th century writer Lactantius: 'a poor mathematician (the term used for astronomer), though in other respects a worthy author, who writes very childishly about the shape of the earth when he scoffs at those who affirm it to be a globe'. Is it not the plain sense of readings in Genesis, Isaiah and the Psalms, Lactantius would say, that the universe is like a house with the earth as its ground floor? Galileo was familiar, too, with Kepler's defence of Scriptural authority *rightly understood.* As a devout Catholic, Galileo wrote:

> (I)t is very pious to say and prudent to affirm that the holy Bible can never speak untruth whenever its true meaning is understood. But I believe nobody will deny that it is very often abstruse, and may say things which are quite different from what its bare words signify. Hence in expounding the Bible if one were always to confine oneself to the unadorned grammatical meaning, one might fall into error.[71]

The Bible, he argued, was written for ordinary unlearned people. Obviously there could be no assertion that the earth moved round the sun, for they would not have believed it and would, no doubt, have begun to question the reliability of Scripture as a whole. The Bible was not intended for scientific enlightenment but for 'the service of God and the salvation of souls'. Quoting a 16th century historian, he wrote that the purpose of the Scriptures was to teach us 'how to go to heaven, not how the heavens go'. The Council of Trent (1545-63) had ruled that the Church Fathers were to be the arbiters in matters biblical, but Galileo concluded that this applied only to matters of faith and morals.

At this stage the official opposition to Copernicanism began to harden. *De Revolutionibus* continued to be approved as an aid to astronomical calculations but not as an account of physical reality. Passages which implied a reinterpretation of the Scriptures were censored. Indeed, the theologians of the Congregation of the Index declared the heliocentric idea 'foolish and absurd, philosophically and formerly heretical ... (it) contradicts the doctrine of Holy Scripture in many passages, both in their literal meaning and according to their general interpretation by the Fathers and Doctors'.

The Pope, Paul V, instructed Bellarmine to summon Galileo and order him not to hold or defend those ideas. This was a bitter moment for

Galileo. How could a person be expected to deny his own senses and subject them to the rule of another? Perhaps he had been confident that, even acting alone, he could persuade the authorities of the Church – otherwise here lay the way to the ruin of the state.[72] Nevertheless it was a moment of truce. No penalty had been imposed on Galileo but simply the prohibition on the promoting of Copernican teaching. For seven years he complied with it.

Then came the election of a more liberal Pope, Urban VIII, who was well predisposed towards science and the arts and commended Galileo for the latter's book on comets. Such openmindedness spurred Galileo on to a renewed pro-Copernican campaign. Writing in the form of a fictitious conversation between a traditionalist (*Simplicio*), a newly enlightened natural philosopher (*Salviati*, who speaks for Galileo himself) and an intelligent questioner (*Sagredo*), he produced his fateful book *Dialogue concerning the Two Principal Systems of the World* (those of Ptolemy and Copernicus). When it appeared in 1632, Galileo's enemies were outraged by such blatant pro-Copernican propaganda. Rome's Jesuits, who had earlier been ridiculed by Galileo for their pro-Aristotelian world-view, now pronounced the *Dialogue* as more pernicious for the Church than all the works of Luther and Calvin. The book ended with the Pope's views being voiced rather tritely and simple-mindedly by Simplicio as he gives his verdict on the entire lengthy debate. This careless piece of satire was evidently the final straw. The Pope's firm but generous attitude changed to anger and in early 1633 Galileo was summonsed to Rome. Aged seventy and in poor health he made the journey, was arrested and appeared before the Inquisition.

Galileo was accused of disobeying the injunction delivered to him in person by Cardinal Bellarmine in 1616. This had been recorded on an unsigned certificate that was now brought forward as evidence of that comprehensive ban. Galileo countered this with his own certificate of the encounter, signed by Bellarmine and describing a much milder injunction.

The hearings took place over a two month period and a plea bargain seemed at one stage to have been agreed – lenient treatment if he admitted his errors – but it was not to be. With perhaps growing enmity from the Pope, Galileo was interrogated under threat of torture, illegal though this was in the case of anyone of his age and therefore most unlikely to be carried out. He was required formally to abandon the Copernican system and to state, however untruthfully, that he accepted the Ptolemaic system – and two days later he knelt in a white penitential shirt to be sentenced to life imprisonment.[73] Ultimately this meant house arrest in his villa near Florence where he stayed until he died in 1642.

The Copernican doctrine was never declared heretical and, in fact, Galileo was tried not for heresy but for disobeying the 1616 injunction, whichever version is considered. Even if the Galileo affair has often been caricatured as a paradigm case of the warfare between science and religion, the historical reality was much more complex and fascinating than such a polarity would suggest. 'Galileo can no longer be portrayed as the harbinger of truth and enlightenment who was pitted against reactionary priests who refused to look through the telescope .... his censure resulted partly from his mishandling of a sensitive diplomatic situation'.[74] On the other hand in the total humiliation that it demanded of Galileo the Catholic Church has been accused of extreme insensitivity. Partly as a consequence of that persecution the centre of creative science moved northward to the Protestant countries, notably the Netherlands and England.

> The cause of Galileo's trial really was Copernicanism, rightly seen as a threat to a world-picture which had existed for centuries. The crisis became so acute because of the intimate association of that world picture with official Catholicism. For the Jesuits the collapse of Aristotelian philosophy meant the destruction of the Tridentine faith.[75] What was at stake was the right of a natural philosopher to pursue his scientific research independently of authority. Galileo certainly wanted that, but without any damage to the Catholic faith.[76]

For over a century Catholic natural philosophers felt obliged to present their astronomical and cosmological treatises in the guise of a mathematical hypothesis. Not until 1757 was the anti-Copernican decree revoked by Pope Benedict XIV. However, Galileo's *Dialogue* remained on the Index of prohibited books until 1831. Some official redress was also made by Pope Leo XIII in his 1893 encyclical which endorsed the approach to the Scriptures taken by Galileo. Three hundred and fifty years after the death of Galileo the present Pope, John Paul II, announced that the Church had erred in condemning him.[77]

The drama of Galileo's confrontation with church authorities can overshadow the scientific work that he carried out before and during his years of house arrest. His early astronomical observations were a major contribution towards the overthrow of the Aristotelian world-view, but his most significant contribution to the Scientific Revolution was yet to come. Unlike most of his fellow academics he was familiar with the practical world of skilled artisans and engineers, which no doubt inclined him towards the recently emerged tradition of applying mathematics to technological problems.[78] But his genius lay particularly in bringing mathematics to bear upon a *limited* aspect of the physical world, namely, the motion of a body falling freely to earth or tracing a curved path as a projectile – *isolating* such motion from any metaphysical ideas, such as Aristotelian final causes, and *idealizing* it by imagining it to occur with as few perturbing influences as possible, such as the frictional drag of the air. The building of scientific understanding as the accumulation of many such piecemeal investigations into a growing framework is taken for granted in modern science – but under the dominance of Aristotelian physics this first such investigation must surely have required a great leap of mathematically informed imagination.

Such idealized localization of the phenomenon, with its exclusion of any Aristotelian final cause, had been the hallmark of Archimedes and his Alexandrian followers (during the third century BCE) in the five disciplines of astronomy, optics, statics (including hydrostatics), mathematics, and 'harmonics' (of vibrating strings). Galileo now extended this approach to the study of motion (kinematics). An important element in his work was that he did not experiment in order to infer a law of nature, but in order to verify *a posteriori* a relation he had deduced by mathematical reasoning from suppositions that appeared more or less self-evident. Or again, his approach was to describe the phenomena of nature in ever more embracing schemata by means of the exact language of mathematics.[79]

Galileo's kinematics was perhaps the key element in turning science from the 'more-or-lessness' of the 16th century into the exactness of Newtonian physics and its associated mechanistic world-view.

> Though there were men in the later middle ages who were carefully observing nature, and improving greatly in the accuracy of their

> observations, these tended to compile encyclopaedias of purely descriptive matter. When there was anything that needed to be explained these men would not elicit their theories from the observations themselves – they would still draw on that whole (Aristotelian) system of explanation .... [80]

The rediscovery and application of the Archimedean approach, from Galileo onwards, meant a turning to a far more empirical approach, epitomised in the vigorous promotion of experimental science by Galileo's contemporary, Francis Bacon. That was one stream of development. The other lay in a restructuring of knowledge in the metaphysical system of Descartes.

FRANCIS BACON (1561-1626)

The names of Francis Bacon and Rene Descartes are often linked as important complementary figures in 17th century science, with their different emphases on scientific method. Whereas Descartes held that it should be possible to discover scientific knowledge primarily by *deductive* thought from general principles, with experiment playing an auxiliary role, Bacon helped to form the scientific enterprise as the collecting of materials, the carrying out of experiments and the *inductive* discovering of general features and principles therefrom.

Bacon entered Cambridge University at the age of twelve with prococious intellectual gifts, and left at the age of fifteen to train as a lawyer and prepare for a career in state administration. He rose through the legal ranks, eventually becoming Lord Chancellor, and was knighted by King James I. But his career ended in shame in 1621 when he was accused of taking bribes from criminals seeking acquittal. In the remaining few years of his life he threw himself into his great project of developing a new philosophy of science.

Immanuel Kant has described Francis Bacon as the central figure in the establishing of the empirical sciences. Scientific observations had previously tended to be arbitrary and unorganized – groping in the dark, as Kant put it. Although not an experimenter himself, Bacon urged and inspired others to organize experiments systematically in order to interrogate nature and discover its laws. He has also been described as a shrewd lawyer who was always at the centre of public affairs, and as 'that great diplomat for science' whose concept of the concerted organization of science led directly to the formation of the Royal Society.[81]

Showing a keen proactive approach to the quest for scientific knowledge, Bacon wrote:

> Instead of this speculative philosophy that is taught in the schools, one can find a practical philosophy by which, knowing the force and action of fire, water, air, the stars ... as distinctly as we know the different trades of our craftsmen, we could employ them in the same way to all uses for which they are appropriate and thus become the masters and possessors of nature.

For him the true and lawful end of the sciences is that human life be enriched by new discoveries and powers and by the discovery of 'new arts, endowments and commodities' for the bettering of man's life. He could declare that he was not labouring to lay the foundation of any sect or doctrine, but of human utility and power.[82] Nowadays such sentiments are easily associated with the darker side of science and technology, the exploitation of nature. The extent of that exploitation could hardly have been foreseen by Bacon and, given the Renaissance sense of liberation and its bursting forth of new life and vitality, his stance seems unsurprising. But the language and imagery he employed has drawn severe feminist criticism.

A recent analysis by Carolyn Merchant focuses on the two coupled assertions: firstly, the Scientific Revolution signified the death of nature, bringing about the change from an organismic to a mechanistic world-view, and secondly, our environmental problems can be traced to the emergence of an image of nature as 'a passive woman to be subdued rather than as a nurturing mother to be revered'. The underlying assumption is that nature has always been metaphorically compared to a woman, with connotations of both nurturing and dominion. In much of the history of Christian thought before the 17th century it was the former that prevailed. With the onset of commercial capitalism, however, the need to draw upon the resources of nature – through mining and other counter-natural activities – steadily increased and the values associated with the nurture view soon took second place.[83]

Francis Bacon is regarded as the crown witness for the upgrading of the 'dominion' image. He is accused by Carolyn Merchant of using language which ultimately legitimates the exploitation and rape of nature for the sake of human good. For example he wrote: 'I am come in very truth leading to you nature with all her children to bind her to your service and make her your slave'; and then, 'You have but to follow and, as it were, hound nature in her wanderings, and you will be able when you like, to lead and drive her afterwards to the same place again'.

There is perhaps no one who has done more than Bacon to shape the ethos of modern science, especially through his vision of what a scientific community might be and do. In his remarkable story *The New Atlantis* (1627),[84] he pictures a marvellous land in which the citizens live ideal Christian lives in an ideal society, made possible by a monastic-like scientific establishment known as Solomon's House. It is a highly patriarchal society, led by a group of thirty-six 'fathers' whose self-appointed mission is to find 'the knowledge of causes and secret motions of things' and to use such knowledge for 'the enlarging of the bounds of human empire, to the effecting of all things possible'.

The story contains a list of things possible from the work of these fathers – artificial metals, gems, stones and other wonderful new materials; flying machines, submarines and machines of war; a huge range of medicines; marvellous optical instruments, including telescopes and microscopes, long before they were invented; the simulation of atmospheric phenomena and even the fire of sun and stars; and techniques to predict natural disasters such as earthquakes, floods, diseases and plagues. Experiments are envisaged on living creatures, such as cross-breeding and the testing of drugs and other chemicals, and on the simulation of the weather in vast caverns in the earth. It was a vision that was amazing in its sheer scale and imagination, stemming from Bacon's belief that humanity could win back the state of grace it had possessed in the Garden of Eden before the Fall. Science would 'save' us by giving us the power to make all lives happy, healthy and comfortable.

As with most natural philosophers in the Scientific Revolution, Bacon's scientific ideas lay within a framework of religious belief. Like Robert Boyle after him, he warned against the mixing of science and religion, conscious no doubt of unwelcome aspects of Renaissance magic with its ideas of hidden natural powers in the cosmos. Yet he remained convinced that scientific conclusions had still to be limited by religious belief. On issues such as the size and eternity of the universe, his faith played a selective role, setting the conditions for admissible theories. An eternal universe, for example, was incompatible with a created one. Like Galileo, his contemporary, he firmly rejected any Aristotelian notion of final causes, which he described as 'barren virgins'. It was essential to concentrate simply on the immediate physical causes of natural phenomena, otherwise science would be impossible.

RENE DESCARTES (1596-1650)

Descartes was an intellectual who made significant contributions in the natural sciences of his day, but it was as a philosopher that he was, and has continued to be, both influential and controversial.

His schooling in France, under Jesuit teachers, included classics, poetry, mathematics and the works of Aristotle. He disliked the scholastic approach[85] that dominated the syllabus and felt it was only the mathematics that helped him in his later work. As a young man he was financially independent and thus able to spend several years on travel in Europe, during which he reflected on a wide range of mathematical problems and developed an enthusiasm for scientific questions. It seems that at the age of twenty-three there dawned on him the possibility of drawing together within the unifying framework of mathematics such distinct sciences as arithmetic, geometry, music, astronomy, optics, mechanics and others. This culminated in a vision which he understood to be a divine revelation of what his central task was to be – a vision of the unfolding of a new and wonderful science.[86]

For Descartes the essential task was to restructure the entire knowledge of reality on philosophical foundations so secure that that which could be said to be known would be known with absolute certainty – that at least was his aim.

He had before him Thomas Aquinas' *Summa Theologiae*, perhaps the greatest of all works of metaphysics, together with the contributions of Copernicus, Kepler and Galileo. But spurning all this earlier work he chose rather to start from scratch and construct his own theistically based account of the natural world. In contrast to Galileo's piecemeal approach, his own can perhaps be described as showing 'the arrogance of philosophical thought which, in one leap of genius, wants to embrace the entire world and aims at once for an insight into the essence of things ...'[87] He has indeed been referred to as the founder of modern metaphysics. Perhaps he saw that the task of completing science had been assigned to his own generation or, more precisely, to Rene Descartes himself.

Descartes claimed a high position for revealed truths, for the way to heaven, he believed, is no less open to the most ignorant as to the most erudite. However, he rejected much of traditional theology, especially that which was tied to an Aristotelian natural philosophy, for his concern was to eliminate all reasoning from theology and see it simply as a body of supernaturally given doctrines. Yet he could write: Now I regard that all whom God has given the use of reason are under obligation to employ it principally to strive to know Him and to know themselves. The main point here is that he made a distinction between the givenness of divine revelation and that part of God's nature that was accessible to the intellect.[88]

It was Descartes' metaphysical ideas about God that formed the foundations of his natural philosophy. In his *Meditations on First Philosophy* (1641) he defined God as an infinite, eternal, immutable, independent, omniscient, omnipotent substance by which all existing things have been created and produced – a definition best summarized by the word 'perfection'. No doubt the contemporary natural philosopher Blaise Pascal had this work in mind when he remarked on the wide difference between the God of Judaeo-Christian thought and the God of the philosophers.

Descartes' metaphysics begins with the axiom that all statements about the nature of things must be treated with scepticism. He determined to start from scratch, accepting only those statements that he knew intuitively to be certain. The clearest idea that came to him in the very process of doubting was that he knew himself to be a thinking being. He could then make his famous declaration, 'cogito ergo sum' (I think, therefore I am). Beyond that he held a clear and distinct conception of a perfect entity – and he provided a closely reasoned argument that such an idea could not have originated within his imperfect understanding but only from the perfect entity itself, namely God. In his detailed reasoning he, like Socrates and Plato long before him, claimed a connection between *reason* and *being*. The more self-evident a thing is to one's reason, the more certain is its existence.

After these foundational ideas about the existence of himself as a thinking being and the existence of a perfect God, the next clear idea was that of matter – the material of the world. Here he based his certainty of the reliability of his conception of matter on the trust that a benevolent God would not place such a clear simple rigorously determined thought in his mind if it were false.[89] Thus did his metaphysics begin to form the basis of his natural philosophy. For example, he claimed that there could be no vacuum anywhere in the universe since the Creator, perfect as he is, would not stop short of creating something throughout all available space. Not surprisingly, counter-claims were made that natural philosophy was not the preserve of theists alone – it was open to atheists too. Nevertheless, the givenness of theological doctrines simply reflected, for Descartes, the infinite gap between divine and human reason.

Reckoning that he had found a clear way to demonstrate metaphysical truths (in a way more evident even than the demonstrations of geometry), he set about using such truths to produce explanations of natural phenomena. In fact, so confident was he in the superiority of deductively-reasoned knowledge, over that which was empirically acquired (since, for him, all knowledge that comes via the senses is inherently unreliable), that he did not hesitate to launch into one after another of such explanations. Not a few of these turned out to be wrong; for example, those concerning the mechanisms of planetary motion and blood circulation. The great defect of Descartes' cosmology was his propensity to suggest mechanisms without the rigour of mathematical analysis. It was Isaac Newton, a generation later, who was to bring the matching of experimental investigation and mathematical description to a new level of sophistication.

Descartes is often remembered for the notorious dualism that he introduced between matter/body on one hand and mind/spirit on the other. He knew he was up against a possibly intractable problem in trying to account for the nature of living things in terms of mechanical concepts alone – a problem that concerned him throughout his career. As far as human beings were concerned, he made a distinction between *voluntary* and *involuntary* bodily processes, and he believed that the latter were characteristic of all animals but the former were to be found in humans only.

A rational soul existed in man alone and gave him the power of reason and the consequent power to operate through his body. But whereas the body, being material, was characterized by its spatial extension and could in principle be divided into smaller and smaller segments indefinitely, the soul was immaterial, indivisible and without any spatial dimension, belonging entirely to the realm of the spirit. Insofar as the human body performed involuntary functions, it shared the nature of all animals, and these functions could be understood as pure mechanisms, explicable in terms of particles of matter in motion.[90]

In possessing a soul, man has the capacity to think. Descartes discussed the notion of 'soul' at great length and went so far as to postulate its location in the human body, for body and soul were clearly intimately connected. Despite the difficulty in imagining how a non-spatially limited soul could interact with a spatially limited body, he believed that body and soul somehow constituted a unified whole, with the soul acting primarily within the brain, especially within that small organ known as the pineal gland – a reservoir for the finest of blood particles which, like messenger particles, could be sent from various pores in the brain through the nerves and into the muscles.[91]

Descartes thus rejected the Aristotelian belief in a 'vegetative soul' which governed nutrition and growth in *any* living thing, and in a 'sensitive soul' which governed sensation and locomotion in animals and humans. He was especially firm in his rejection of the notion that animals might possess souls. He wrote:

> There is nothing which turns weak spirits sooner from the correct path of virtue than to imagine that the soul of beasts might be like ours, and that consequently we have nothing to fear nor to hope for after this life, any more than flies and ants; while when we know how different they are, we understand far better the reasons which prove that our soul is of a nature entirely independent of the body

> and therefore that it is not subject to die with it; then, seeing no other causes for its destruction, we are naturally led to conclude that it is immortal.[92]

For Descartes it was thus important to make this sharp distinction between animals and human beings. Man's body functioned like that of an animal, but his possession of a soul made him a unique creature, possessor of an immortal soul.

Largely mistaken in his mechanisms and perhaps old-fashioned and over ambitious in his all-encompassing philosophical search, Descartes nevertheless seems to have played an indispensable role in providing the Scientific Revolution with a suitable metaphysical substructure and, together with his two French contemporaries, Marin Mersenne and Pierre Gassendi, laying much of the groundwork for the mechanistic philosophy of the middle and late 17th century.[93] In the decade after Descartes' death this philosophy was taken up by Robert Boyle, Robert Hooke and other prominent members of London's newly established Royal Society, in varying degrees and with different emphases.

The appeal to mechanism could be regarded in different ways. For some it was seen as a new concept of causality which affirmed action through mechanical contact since, with Descartes, they tended to be suspicious of any notion of action-at-a-distance with its psychic connotations. For others, especially Boyle, it represented mainly a new theory of matter in which the properties and qualities of any type of matter were simply ascribed to the configuration and motion of its constituent particles. And for them all, no doubt, it encouraged the explanation of natural phenomena in terms of mechanical models or analogues.[94]

ROBERT BOYLE (1627-1691)

The Honourable Robert Boyle has been described as the father of modern chemistry and a man of remarkable piety – a member of the Church of England who was very much a Puritan at heart. At the age of twenty-one he experienced a deep spiritual crisis in which he resolved to spend the rest of his life in dedication to the service of God. He immersed himself in biblical theology, learning Hebrew in addition to the Greek he already knew because he regarded the truths in the Bible as 'so precious and important that the purchase must at least deserve the price'.

In his early adult years, Boyle was a member of a group in Oxford who were passionate believers in Bacon and his Experimental Philosophy and although he appreciated the importance of mathematics and regretted his lack of it, he believed that the fundamental task in natural philosophy was to broaden its experimental basis – to inquire into the ways of nature *empirically* without being drawn too soon into forming hypotheses. Indeed the contemporary Dutch scientist, Christiaan Huygens, came to remark that in relation to the vast amount of experimental work that Boyle had recorded, he had made very few important discoveries and generalizations.[95] Nevertheless, he was to make his mark in the histories of both chemistry and natural theology.

The most important of his scientific writings was entitled *The Sceptical Chymist*, in which he laid bare the shortcomings of Aristotle's doctrine of 'substantial forms' – that all substances comprise in varying proportions a combination of the four basic elements of *earth, water, air,* and *fire* – and attacked the alchemists' scheme in which all matter is characterized by combinations of the three basic principles of *sulphur*, *salt*, and *mercury*. In this seminal work he helped to lay the foundations of modern chemistry, even if it took another century to place the science of chemistry on its feet.

Despite his reluctance to formulate hypotheses about the processes of nature, Boyle was drawn to the corpuscular view of the universe that had been promoted by Bacon and developed by Gassendi and Descartes. This mechanical philosophy, as he often called it, constituted the framework for his understanding of the universe as mechanical at all levels, from particles to planets, explicable in terms of the three underlying principles of *matter*, *motion*, and *rest*.[96]

Here matter was regarded as capable of being reduced to extremely small particles, indivisible except in one's imagination or by divine power. He envisaged these *prima naturalia* as coalescing into 'primitive concretions or clusters', too small to be observed, seldom broken apart, and sufficiently varied to explain the different characteristics of physical bodies, including their colour, taste, and smell. He suggested that sweetness or sourness depended on the absence or presence of sharp edges in the corpuscular constituents of the material being tasted. Or again, matter was fluid when its corpuscles touched one another over a relatively small surface area, easily gliding along each other.

The understanding of the universe as mechanism formed an integral part of Boyle's religious outlook. Indeed he made a profound synthesis of science and faith, seeing his scientific work as a matter of religious obligation and even as a form of worship, requiring purification from earthly desires. His piety included a vow of lifelong chastity, an attitude that was not uncommon among the early Fellows of the Royal Society and other *virtuosi* (the name by which the English scientists of the late 1600s were known), many of whom were numbered among the Puritans of English society.[97]

What was the nature of the English Puritanism of which Boyle was so notable an example? One aspect of the Puritan approach was its encouragement of the new philosophy – the new anti-Aristotelian science based on 'manual experiments', with Francis Bacon as its prophet. Puritan piety and the Baconian experimental philosophy had much in common: anti-authoritarianism, optimism about human possibilities, and a vigorous rational empiricism. It is an absurd idea, wrote C S Lewis, that the Puritans were somehow grotesque elderly people, outside the main current of life. In their day they were the avant-guarde. Unless we can imagine the freshness, the audacity, and (soon) the fashionableness of Calvinism, he wrote, we shall get our whole picture wrong.[98]

Despite the link in that period between Puritanism and early modern science, Boyle felt it necessary to defend his synthesis of science and Christian belief. Among his apologetic writings is an essay on *The Christian Virtuoso*,[99] explaining how it is that 'a great esteem of (scientific) experience and a high veneration for religion should be compatible in the same person'. In this work Boyle was especially concerned to respond to the great and deplorable growth of irreligion, shown in the lives of those he referred to as 'these infidels and libertines'. The motivation of his work was, above all, to convince others of the excellence of the creation and thence the goodness and wisdom of its Creator. He was particularly impressed by the excellent structure of animals and plants, especially the structure and function of the eye in different animals. A rational person could not but conclude that 'this vast, beautiful, orderly, ... admirable system of things that we call the world, was framed by an author supremely powerful, wise, and good'.

In another work entitled *A Disquisition about the Final Causes of Natural Things*, he asserts the importance of the search for final causes, that is, the *ends* for which the constituent items and systems in nature show design. Here he is critical of Descartes who, in establishing his anti-Aristotelian position, excluded all final causes from his world-view, allowing God simply the role of First Cause. Thus did Descartes throw away well-tried argument, wrote Boyle, shown to be successful in establishing among philosophers belief in and veneration of God.

Again, where Descartes thought it presumptuous of man to pretend to discern the purposes of God – of a Mind infinitely beyond the reach of any human intellect – Boyle would say that some of those purposes could indeed be known, in order that a man could give the praise he owed his Creator. 'God may, if he pleases, declare truths to men, and instruct them, by his creatures and his actions, as well as by his words'. Whereas Descartes' metaphysics was regarded by some as dangerously supportive of atheism, Boyle defended that scheme as resting on the assumption of God's existence. Individuals become atheists, claimed Boyle, not through the influence of the mechanical philosophy but from their own immorality.[100]

However, the mechanical philosophy raised afresh the question of the nature and extent of divine activity in the world. Nature is like a rare clock, wrote Boyle, one vast interlocking mechanism such as that at Strasbourg. But if this analogy was drawn too far, it became difficult to see what room there could be for the Creator to act. For Boyle this complication was worth enduring if it enabled one to claim the sovereignty of God as the source and sustainer of a superbly constructed world. Furthermore, he speculated (as did others before and after him) about the possible analogy between the action of the human spirit within the human body and divine action within physical process.

Boyle was sufficiently aligned with Christian orthodoxy to believe that Christ and his disciples performed miracles – special events which breeched the laws of nature – as demonstrations that the power of God has intervened during the course of history and as signs of the truth of Christ's teaching. Yet almost the entire emphasis in his apologetic writings was on the power and wisdom of God as Creator. It was not that he experienced any particular pressure against the plausibility of orthodox teaching on the divine work of incarnation and salvation, but rather that the workings of nature were now so brilliantly illuminated that the doctrine of creation seemed the obvious route to explaining mankind's situation. Conscious, too, of how the heavens declare the glory of God, the *virtuosi* saw the divine handiwork in ways not possible to any previous generation, and their writings express awed surprise and admiration for the skill of the Creator.[101]

The natural theology that grew in late 17th century England, and continued well into the 19th century, concentrated on the argument for the existence of a beneficent God from the evidences of design in nature. In the early years of this development, Boyle was a key figure, not only for his direct influence but also through his endowment of the Boyle Lectures. These were intended, he wrote, for proving the Christian religion against notorious infidels, (that is) atheists, theists (deists), pagans, Jews, Mahometans, not descending lower to any controversies that are among Christians themselves. They were to be delivered annually by a member of the clergy. With their pointers to stability, unchanging truth and eternal certainty, they met a need in the restless English society of that time and strengthened the programme for Christian apologetics espoused by their founder. They soon became very influential and enjoyed continuing success through their frequent reprintings and the many tributes to them by men of letters.[102]

Richard Bentley, a classical scholar, delivered the first Boyle Lecture on 'A confutation of atheism from the origin and frame of the world' after conferring with Newton on scientific and philosophical issues. In his first reply to Bentley, Newton stated:

> When I wrote my treatise about our system (*Principia*) I had an eye upon such principles as might work with considering men for the belief of a Deity and nothing can rejoice me more than to find it useful for that purpose.

Bentley reflected something of Newton's thinking when he urged that universal gravitation exists above all mechanism and material causes, and proceeds from a higher principle, a divine energy and impression – and that the present frame of sun and fixed stars could not possibly subsist without the Providence of that almighty Deity 'who spake the word and they were made'. He claimed the existence of God not only from the evidence of superb design in the cosmos but also from the necessity for its ongoing need for divine maintenance, as argued by Newton himself. Many of the subsequent Boyle lecturers concentrated in various ways on the argument from design, seeking especially to ground their theology in the Newtonian world-picture.

It has been argued that the ultimate purpose of the Boyle Lectures was to underpin both the established Church and the social order, forming the cornerstone of a liberal, tolerant, and highly philosophical version of Christianity, a natural religion based upon reason and (Newtonian) science.[103] It is also strongly counter-argued that there is little unambiguous evidence for this sociological judgment. We should assume rather that the primary purpose – however misguided – was to use science to defend a faith under threat.[104] By the time the Lectures were launched, Newton's ground-breaking scientific work had been published, giving a great boost to the Enlightenment idea of 'man the measure of all things'.

ISAAC NEWTON (1642-1727)

Sir Isaac Newton is best known as a scientific genius whose monumental work, *Mathematical Principles of Natural Philosophy* (1687) – commonly known as *Principia Mathematica*, or simply 'the *Principia*' – has been described as perhaps the greatest intellectual feat in all of science.[105] The *Principia* covers the following topics in its three sections: firstly, the general principles governing the motion of bodies (that is, *dynamics*); secondly, the application of these principles to the motion of celestial bodies, especially the planets and their satellites, under the force of gravity; and thirdly, fluid mechanics, the theory of waves and certain other aspects of general physics, all grouped together in one section.

However, his philosophical interests ranged further afield, embracing theology and the arcane pursuits of alchemy at least as much as natural philosophy.[106]

Newton came from a successful farming family in Lincolnshire where, from the age of three, he was brought up by his grandmother, after the death of his father and subsequent remarriage of his mother. As a boy he built models of windmills and waterwheels and filled the house with his homemade sundials, becoming adept at telling the time simply from the positions of the shadows in the house. Indeed, from an early age he had been fascinated by the sun and the nature of its light, and by the daily and yearly variations in the positions of shadows.[107]

All his life Newton remained a very reserved and retiring person, unmarried, although in later years he became much involved in the affairs of the Royal Mint (as Master) and the Royal Society (of which he became President). After grammar school he entered Trinity College, Cambridge, where Aristotelian physics was still being taught. He had to find his own way to the works of Kepler, Galileo and Descartes, which launched him into the physics of motion. At the age of twenty-six he was made Professor of Mathematics at Cambridge and was widely regarded as Europe's foremost mathematician. In due course, with the publication of the *Principia*, he became its leading physicist too. He was elected President of the Royal Society in 1703 and was knighted in 1705.

Newton's *Principia* marks the climax of the Scientific Revolution. In it lay the convergence of two intertwined streams of thought: the *Platonic-Pythagorean* tradition which described nature in geometric terms, believing the cosmos to be constructed according to principles of mathematical order; and the *Democritean* tradition or *mechanical philosophy* in which nature is perceived as a vast machine and which sought, by empirical investigation, to explain the causal mechanisms of the phenomena of nature – phenomena essentially reducible to the movements of particles of different shapes and sizes.

During the great plague of 1665-6 Newton took the precaution of secluding himself in his Lincolnshire home for almost two years. It was a period of immense creativity in which, having mastered the essentials of the existing mathematics, he developed the powerful tool of differential calculus from the work of Fermat and Descartes (although the question of priority in this crucial advance was bitterly contested by Leibniz), discovered certain other mathematical techniques, produced a theory of the spectrum of colours in white light, and began to think of gravity extending to the orbit of the moon.

In due course Newton developed Galileo's mathematical relations for freefall motion into his own famous Laws of Motion:

- A body will continue to move with constant velocity unless compelled to change its velocity by an applied force (the law of inertia).
- The change of velocity is proportional to the force applied and is parallel to it (or 'force equals mass times acceleration').
- To every action (force) there is an equal and opposite reaction (force).

Combining the second of these with Kepler's laws of planetary motion, he formulated his celebrated Law of Universal Gravitation for the force F between two masses, M_1 and M_2, separated by a distance R:

- F is proportional to the product M_1M_2 and inversely proportional to R squared.

He derived this firstly for the force between the sun and any planet and then, by remarkable intuition, postulated that the equation represents the force between *any* pair of masses in the universe. It is the same universal force, he claimed, that keeps the planets and their satellites in their respective orbits and causes objects to fall towards the earth – and, indeed, is responsible for the ocean tides.

Newtonian physics is the basis of much of modern technology, including the complicated enterprise of sending spacecraft on planetary missions. To a historian, however, the importance of Newton lies primarily in his provision of the first full explanation of the universe on mechanical principles, expressed in terms of the three laws of motion and the law of universal gravitation applied to all matter everywhere, on earth and in the heavens. Descartes' vision of nature as a perfectly ordered machine governed by mathematical laws was thus fulfilled, even if it involved the bold idea of force acting between well separated bodies without any mechanical connection. Perhaps this idea came to Newton through his deep involvement in alchemy, an all-embracing philosophy that men of his day found immensely attractive.

> If it resembled the mechanical philosophy in its scope, alchemy parted from it decidedly on a question of importance to most men of (that) century. Without denying God as Creator, the mechanical philosophy denied the participation of spirit in the continuing operation of nature. Alchemy did not merely assert the participation of spirit; it asserted the primacy of spirit. All that happens in nature is the work of active principles, which passive matter serves as a mere vehicle.[108]

As the 1680s progressed, the role of what the alchemists called 'active principles' took on far greater importance to Newton and led him to a radical assessment of how gravity operated. This change in perception was perhaps the most important step in Newton's development of universal gravitation. The authoritative Newton scholar, Richard Westfall, has gone so far as to say that Newton could not have visualised attraction at a distance had it not been for his alchemical work.[109]

Committed mechanists like Galileo and Descartes would have criticised this key assumption because it lay uncomfortably close to the still widely held belief in occult influences of celestial bodies. The notion of some sort of celestial force had been suggested previously by Kepler who took a serious interest in astrology, and it is perhaps significant that Newton wrote extensively in the area of hermetic philosophy and alchemy[110] – an aspect of his life's work about which he was very secretive. Indeed, given the rising acceptance of the mechanistic view of nature as a despiritualised world of hard matter, astrological interests tended to be kept private. Boyle, for example did not allow his defence of astrology to be published until after his death, and Newton and his literary executors sought to suppress the hermetic background of Newton's scientific ideas with its arcane imagery.[111]

However, for occult believers and sceptics alike, it was the elegance and universality of Newton's account of motion and gravitation, together with its congruence with observation, that soon made it thoroughly convincing – to those, at least, who could follow it. For Newton himself, the forces of gravity, however puzzling, were perceived as 'puppet strings' by which the Creator maintained the regularities of the celestial motions.[112]

Apart from his hermetic interests, Newton showed strong and independent religious belief, a well known aspect of his life which some commentators have sought to side-step, or at least see as unlinked to his science. Others have argued that he considered his new system of the world to be a weapon against the growing unbelief of that period and that he wrote the *Principia* with its apologetic value in mind. In any case, it is clear that religious belief was something quite basic to him apart from its link with his science. Underlying it was evidently some sort of religious experience, detachable from his theological framework, which reinforced the conviction that the world of science is by no means the whole world.[113]

Newton was also a prolific writer on theological topics. Although he may have overestimated the importance of his biblical and theological scholarship, he was once described as 'knowing more of the Scriptures than them all', and his religious beliefs certainly affected the way he understood and interpreted the physical world. Furthermore, he showed a keen interest in biblical prophecies, not for any predictive power, for he believed them to have been fulfilled in or before his time. Indeed, he maintained that the end of prophecy is not to make us prophets but rather to proclaim the wonder of fulfilled prophecy.[114]

Privately Newton rejected the formal doctrine of the Trinity since he did not find it exhibited explicitly in the New Testament. However he could still write:

> There is one God, the Father, ever living, omnipresent, omniscient, almighty, the maker of heaven and earth, and one Mediator between God and man, the man Christ Jesus ... (W)e are to worship the Father alone as God Almighty, and Jesus alone as the Lord, the Messiah, the Great King, the Lamb of God who was slain, and hath redeemed us with his blood, and made us kings and priests.[115]

From his theological writings as a whole he has been described as a pious, believing Christian in all that the term then implied. But he held Arian ideas about the person of Christ (as a being intermediate between God and man, therefore less than fully divine) and knew that he could not make the declaration of adherance to the Church's doctrine of the Trinity and therefore could not be ordained as a priest, the customary requirement for a fellowship at an Oxford or Cambridge college. He expected to be obliged to leave the university, but at the last moment was granted a royal dispensation to become a fellow without being ordained.

Within the *Principia* itself Newton made the assertion that God placed the planets at different distances from the sun so that according to their degrees of density they may enjoy a greater or less proportion of the sun's heat – a direct expression of what might be called his theology of nature.[116] His experience of the natural world and the writings of earlier natural philosophers had convinced him of the reality of intelligent purpose in the cosmic order. Does it not appear, he wrote, that underlying the phenomena of nature there is a being incorporeal, living, intelligent, omnipresent who, in infinite space – as it were in his sensory – sees the things themselves intimately? In fact he regarded the main purpose of natural philosophy as the search for the First Cause and thence to resolve questions such as the following: How does the force of gravity occur? What stops the stars from falling towards each other? What is the significance of comets, and why is their behaviour so different from that of the planets? How is it that organic life is so excellently functional? What is the source of all the order, economy and beauty in the world?

Of all the teleological arguments, however, the most compelling for Newton lay in a certain aspect of celestial structure: the fact that 'planets move all one and the same way in orbs concentric, while comets move all manner of ways in orbs very eccentric'. It seems that he ascribed such waywardness in the motions of the comets to some natural cause, whereas planetary motions 'must have been the effect of counsel'. No *natural* cause, he maintained, could have given the planets and moons the precise velocity that each needed to maintain an almost circular orbit (rather than a hyperbolic, parabolic or eccentric elliptical orbit). To get all these motions right in so great a variety of bodies argues that the cause of it all 'be not blind or fortuitous but very well skilled in mechanics and geometry'.

Newton refers at times to natural processes being governed by active principles which he invokes as agents of God by which the total amount of motion in the universe is conserved; without their agency the motion would steadily be dissipated. At other times he implies that God, as the consummate mathematical mechanic, is directly active in the cosmos – for example, enabling the forces of gravity (the 'puppet strings of God') both to maintain the orbits of the planets and to nudge them back on course whenever their accumulation of interactions with other celestial bodies leads them towards instability. (It was at this point that Newton was accused by his great rival, Leibniz, of harbouring the unworthy idea of God as an inefficient creator who has to keep tinkering with his creation to get it right). Consequently, where others had drawn the conclusion, from the dramatic success of Newtonian physics, that gravity is an innate property of material bodies, Newton himself wrote vehemently that such an idea was 'so great an absurdity that I believe no man who has in philosophical matters a competent faculty of thinking can ever fall into it'.

Newton's thinking in the realm of physics was deeply influenced by his framework of religious belief – as indeed were key figures in the development of geology more than a century later. Because of his belief that the Creator intervenes intermittently in the world's natural processes, especially the planetary motions, he has been labelled – like Boyle and some of the geologists of the early 19th century – a *semi-deist*.

He lived through a period in which the Western world-view was changing rapidly. This was soon to take on the note of enthusiastic confidence in the power of human reason that characterized the Enlightenment of the first half of the 18th century.

THE SIGNIFICANCE OF THE SCIENTIFIC REVOLUTION

The Scientific Revolution was in some ways the final expression of the Renaissance – that rebirth of culture that has been described as the most amazing bursting forth of new life and vitality the West had seen for a thousand years.[117] For example, it has been persuasively argued that there is a vital connection between the tradition of Hermetic magic in the Renaissance and the emergence of early modern science.[118] Nevertheless, the latter ushered in an essentially new and very different way of thinking about the world, contributing to the overall sense of rapid change that was a central feature of the late 17th century.

> For two thousand years the general appearance of the world and the activities of men had varied astonishingly little ... so much so that men were not conscious of either progress or process in history .... Now, however, change became so quick as to be perceptible with the naked eye, and the face of the earth and the activities of men were to alter more in a century than they had previously done in a thousand years.[119]

Butterfield describes this period as one of the great episodes in human experience when new things are brought into the world and into history out of men's own creativity, and their own wrestlings with truth. Furthermore, it was part of a more general and complex process of rapid change, combining with other factors to create what we call the modern world. In the earlier decades of the twentieth century historians have often been tempted to think of the modern world as the product of the Renaissance as a whole, failing to realize the radical nature of the changes implicit in the new science itself. Indeed, there emerged from the 17th century

> a kind of Western civilisation which when transmitted to Japan operates on tradition there as it operates on tradition here (in Europe) – dissolving it and having eyes for nothing save a future of brave new worlds. It was a civilization that could cut itself away from the Graeco-Roman heritage in general, away from Christianity itself – only too confident in its power to exist independent of anything of the kind. We know now that what was emerging towards the end of the 17th century was a civilization exhilaratingly new perhaps, but strange as Ninevah and Babylon. That is why, since the rise of Christianity, there is no landmark in history that is worthy to be compared with this.[120]

What that landmark helped to launch was a colossal secularization of thought in every possible realm of ideas, following the extraordinarily strong religious character of much of 17th century thought.

But there were other powerful secularizing factors at work too. The late 17th century emerges as one of the lowest points in the history of Western Christianity between the 11th and 20th centuries, characterized by a general relaxation in religion and morals after a period of great tension and civil strife.[121] Another contributing factor was the transition from an agrarian economy to that of a commerce-orientated nation-state, in which power moved from the clergy and aristocracy to the middle classes. A third factor was the growth of travel, especially voyages of discovery, creating an awareness of distant lands with long-established religions and cultures. Consequently there developed a sense of the relativity of religious beliefs – and of a less absolute Christianity which harmonized with the new science itself and could sympathize with the deism and pantheism which Newtonian physics made plausible.

The outcome of the Scientific Revolution has been described in general terms thus:

> Between the 15th and 17th centuries, the West saw the emergence of a newly self-conscious and autonomous human being – curious about the world, confident in his own judgments, skeptical of orthodoxies, rebellious against authority, responsible for his own beliefs and actions, enamored of the classical past but even more committed to a greater future, proud of his humanity, conscious of his distinctness from nature, aware of his artistic powers as individual creator, assured of his intellectual capacity to comprehend and control nature, and altogether less dependent on an omnipotent God. This emergence of the modern mind, rooted in the rebellion against the medieval Church and the ancient authorities, and yet dependent upon and developing from both these matrices, took the three distinct and dialectically related forms of the Renaissance, the Reformation, and the Scientific Revolution ... . Out of that profound cultural transformation, science emerged as the West's new faith.[122]

By the middle of the 18th century, western Europe was experiencing a widespread sense of liberating knowledge that its writers described as 'enlightenment'. Light had dawned and what had been obscure was now clear. That feeling was vividly expressed by the French writer Jean D'Alembert (1717–1783):

> If one examines carefully the mid-point of the century in which we live, the events which excite us, or at any rate occupy our minds, our customs, our achievements, and even our diversions, it is difficult not to see that in some respects a very remarkable change in our ideas is taking place, a change whose rapidity seems to promise an even greater transformation to come. Natural science from day to day accumulates new riches. Geometry, by extending its limits, has borne its torch into the regions of physical science which lay nearest at hand. The true system of the world has been recognized, developed , and perfected ... In short, from earth to Saturn, from the history of the heavens to that of insects, natural philosophy has been revolutionized; and nearly all other fields of knowledge have assumed new forms .... Spreading through nature in all directions like a river which has burst its dams, this fermentation has swept with a sort of violence everything along with it which stood in its way.[123]

The rise of modern science challenged the Church of the 17th and 18th centuries, especially in its Protestant form, to reconsider its ways of understanding the world and its Creator. The English *virtuosi* in particular had expressed reverence and awe before the majestic handiwork of God but the big question was (and always is) Who is God? What kind of God is it with whom we have to deal? Some decades previously, Galileo and others had spoken of the book of Nature ('the works of God') as complementary to the book of Scripture ('the words of God'), but a general mood of the Enlightenment was to downgrade the authority of the past, including that of the biblical writers.[124] Consequently, given both the cultural distance of the latter and the huge impact of the new world-picture, the Church was somewhat torn between its commitment to preserve the faith 'once and for all delivered to the saints' and the need to adapt its apologetics to the new age of science.

Some thinkers were drawn to the idea of a reasonable and universal religious faith, open to all people and independent of any particular historical revelation. Its central beliefs lay in the existence of a Supreme Being (often referred to as Beneficent Designer, All-Wise-Creator and other rather impersonal names), the immortality of the soul, and the obligation to moral conduct. Many of the *virtuosi* believed that these constituted the essence of Christianity. Given their admiration for the created order,

> religion was becoming less a matter of living experience than of intellectual demonstration. Nature, not history, was the clue to the knowledge of God. God as Creator, not God as Redeemer, was now the focus of interest. God's relation to the individual, and man's experience of forgiveness and reorientation, were seldom mentioned.[125]

In the mid-17th century Blaise Pascal had asserted 'I believe in the God of Abraham, Isaac and Jacob, not in the God of the philosophers'. In the next century reason largely replaced revelation as the way of knowing God. This dramatic change in Christian thought from within its own ranks prompted the Anglican bishop and historian, Stephen Neill, to describe it as one of the gravest periods of peril through which the Church, in all its long history, has been called to pass.[126] This was the background against which scientific thought and activity continued in a lower key through the 18th century, even as the Enlightenment continued to burn brightly.

3 NATURAL HISTORY: FROM LINNAEUS TO DARWIN

EARLY NATURAL HISTORY AND DEISM

During the late 18th and early 19th centuries, fresh ideas of earth history emerged from the new historical sciences of geology and biology. These culminated in Darwin's theory of evolution which challenged traditional beliefs about humanity's place in nature and, like the Copernican picture, raised the fundamental question of biblical exegesis in the presence of new knowledge. This question recurred all the more acutely since the controversy was not merely about a way of perceiving the world (as introduced by Copernicus) but about the historicity of the biblical creation and flood. At least as important for theology, besides the interpretation of Scripture, was the question of the role of the Creator in the development and history of the world and its creatures.

Speculation about the origin and development of the earth has a long history, much of it concerned with the occurrence of violent earthquakes and floods. Natural philosophers in late 17th century England had written about these from within a largely biblical world-view. William Whiston (1667-1752), who succeeded Newton as Lucasian Professor of Mathematics at Cambridge, wrote *New Theory of the Earth* (1696) in which he identified the force of gravity with God's 'general, immechanical, immediate power'. He was excited by the realization that the Genesis flood could be confirmed by calculations showing that a particular comet, actually seen by himself, would have been in the right place at the right time to have triggered it. He considered that the earth's surface cracked, material from the comet's tail rained down, and the fountains of the deep burst forth from below.[127]

A few years earlier Thomas Burnet had written, in a similar vein, *Sacred Theory of the Earth* (1684). He repeated the warning of St Augustine that it was dangerous to bring Scripture to bear upon disputes about the natural world because advances in science could show Scripture to be wrong and therefore compromised on far more important matters. However, noting that Augustine had used the Bible to dismiss the idea of inhabitants on the opposite side of the world, Burnet proceeded to fall into the same trap: he inferred aspects of the physical world from the Bible. For example, he offered a mechanistic account of how the Genesis flood had come about, as an event in the history of the earth, and he defined the main epochs of that history with reference to information gleaned from the Bible. Perhaps the answer to this apparent inconsistency is to be found in a shifting of the boundary between the domains of science and religious belief over the centuries.[238]

The 18th century brought a notable shift of intellectual concerns. There was a gradual unravelling of the close interaction between science and Christian belief, marked by an increasing note of dissent from biblical orthodoxy about the creation, especially amongst the French *philosophes* who, on the whole, tended to hold to a naturalistic account of the world.[129]

One of their number, the naturalist Benoit de Maillet (1659-1738), produced a speculative world-picture based on the geological investigations he carried out while acting as consul general in Egypt. In his posthumously published *Telliamed* (1748) he criticised the idea of a beginning of the material world (a world of matter in motion) as repugnant to reason. He considered this world to be in a state of development and set forth the history of the earth as a series of epochs: a time when the earth was totally covered by sea; an age of some two million years when the waters receded and the first trees and grass clothed the mountains; then the beginning of animal life, with the early species emerging and evolving from the creatures of the sea, and finally man. Despite its elements of pure imagination it was a picture which influenced the natural philosophers of the Enlightenment and tended to reinforce the picture of nature without God.[130]

With the flowering of the Enlightenment in the mid-18th century the Catholic Church in France seemed solid. Catholics alone had the right of citizenship. Education, the care of the sick, and the institution of marriage were all under ecclesiastical control. But underlying it all there were grave abuses. Furthermore, given the absence of spiritual vitality, there was little concern for the contemplative life, and religion had become an uninspired moralism. All in all it was a languid and lukewarm church, rather than a hopelessly corrupt one, that had to face the ordeal of the Revolution (1789). On the intellectual front it was deeply influenced by the rationalistic and deistic outlook that had emerged in both Europe and North America.[131]

What was the nature of the deism of the Enlightenment? It was characterized above all by an emphasis on the search for the understanding of God through the exercise of reason alone. True religion, they believed, must be universal – a religion that is natural to all humankind – not based on revelation, nor on historical events, but rather on the natural instincts of every human being. At its most basic it stood for the existence of God, the immortality of the soul, and the moral order.[132]

Deists sought to strip Christianity of revelation and they rejected priests, ceremonies, the miraculous suspension of natural laws, superstitions, and any doctrine they regarded as incomprehensible, including the idea of the Trinity and belief in the resurrection of Christ. Also they tended to hold to the notion of God as an absentee landlord who created the world and left it to run itself according to the laws of nature – laws set by divine rationality and accessible to human understanding. They valued the right to think freely and the capacity for independent rational thought.

Although critical of orthodox Christian doctrine they believed in God and thus evoked the hostility of atheists.[133]

In Britain the scepticism of the *philosophes* was paralleled to some extent in the works of the philosophers John Locke and David Hume. However, British society remained more conservative, both socially and intellectually. In particular the occurrence of the biblical flood was generally assumed without question and was invoked time and again to explain the data about rocks and fossils, as such information was accumulated through the 18th century. Yet it was not uncommon in the mid-1700s to hear Anglican clergy bewailing the discrediting of Christianity in a groundswell of rationalism and ridicule.

> Attacks on the power of the Christian churches ... were launched by deists, who denied the authority of doctrines supposedly derived from revelation; by materialists who denied a duality between matter and spirit; and by agnostics who, like David Hume (1711-76), argued that it was impossible to know anything about the nature of God that need affect human conduct.[134]

One of the deists of that period was John Toland (c 1670-1722), an Irish Catholic who turned Protestant. In his book *Christianity not Mysterious* (1696) he set out to show that there is nothing in the Gospel contrary to reason, nor above reason, and that no Christian doctrine can be properly called a mystery. He believed that in its earliest form, Christianity had been rational and intelligible. Men had accepted it not through blind obedience but through understanding it. It had then degenerated into the corrupt complexity of its hierarchical structure, initiation ceremonies, the sign of the cross, images, altars, music, fasts, riches, and pomp, all reminiscent of paganism. For Toland whatever beliefs did not rest on full understanding amounted to no more than superstitions and prejudice. He criticized the 'partisans of mysteries' who fly to miracles as their last resort, yet he believed that even miracles must be possible and intelligible – extraordinary, of course, but possible to God who produces them according to the laws of nature.[135]

Against this background of growing critical thought and dissent, natural history continued to develop. With the growth of travel during the 18th century there came an increased awareness of the huge variety within the world of nature; the collection and classification of plant and animal specimens from many lands grew in scope, and this was followed by a great deal of work in stratigraphy (the study of rock layers) and paleontology (the study of fossils), especially during the first forty years of the 19th century. At that time much of science – especially natural history – was the domain of serious amateurs, many of whom displayed a passion for collecting rock specimens and fossils.

The period 1790-1820 has been called the heroic age of geology when organized observation of the form and content of the earth's crust began to replace the less bridled speculation of earlier generations. It was the first science to be concerned with the reconstruction of the past development of the natural world, especially its rock strata and the fossils they contained. By the introduction of a principle of uniformity – that the processes that have formed the earth throughout its history are those that are observable now – geologists could begin to interpret the past development of the earth in terms of its present state. Such a *uniformitarian* principle immediately led to a greatly lengthened time-scale for the earth's history. On the other hand, the existence of volcanoes and earthquakes, together with indications of widespread deluges, prompted an alternative approach, *catastrophism*, in which major cataclysms were invoked in addition to the more gentle processes. A lively debate continued around the beginning of the 19th century between the proponents of these two schools of thought.

With such attention to the earth's crust and its life forms, two things became clear: that the earth is vastly older than had been imagined (bearing in mind the slowness of the crust-forming processes), and that life has developed from very simple to highly complex forms, ranging from the simplest plants and invertebrates to fish, reptiles, birds and mammals.

We now consider nine of the leading figures and their contributions to the emerging sciences of biology and geology, from roughly the mid-18th to the mid-19th centuries, noting some of their theological concerns. Linnaeus, Buffon, Lamarck and Cuvier each made substantial contributions to the development of natural history, which acquired the name 'biology' in that period. Buffon and Cuvier were also concerned with questions about the age and history of the earth. Hutton, Buckland, Sedgwick and Lyell were primarily devoted to the new science of geology, with Hutton and Lyell as staunch promoters of uniformitarianism and Buckland and Sedgwick making a stand for catastrophism. It fell to Darwin to begin his career in geology and paleontology, and ultimately develop his famous theory of evolution as a wide ranging explanation of the world of nature.

CARL LINNAEUS (1707-78)

Linnaeus was Sweden's greatest naturalist in the 18th century and the father of modern biological taxonomy – the classifying of living things, plant or animal. He identified every organism by two names, the first representing its *genus* and the second its *species*.[136] In his lifetime he achieved wide fame for this simple taxonomic scheme, which is known to be flawed in some of its criteria of classification but is still the starting point for modern versions. His *Systema Naturae* was first published in 1735 and in his lifetime went through another eleven editions of ever increasing length.

In 1732 he journeyed to Lapland, an unhospitable and largely uncharted region, returning with many new species of plants and a great enthusiasm for the task of identifying and classifying them. After a three year spell in Holland (with visits to France and England) he was appointed to a chair of medicine at Uppsala University where he spent the remainder of his career teaching natural history. Students flocked to his lectures. Many travelled widely and brought back botanical specimens. Naturalists from all over the world came to see him and for the last twenty years of his life Uppsala became the centre for European botany. When he died all Sweden mourned, and many regretted the purchase of his fabulous collection of specimens and books by the English naturalist Sir Joseph Banks. This treasure formed the basis on which London's Linnaean Society was founded in 1788, the venue at which Charles Darwin's theory of biological evolution was to be announced.

Linnaeus was a firm believer in the biblical doctrine of the creation. He pictured the original Garden of Eden as a small island near the equator and saw a close parallel between his own work of naming species and the task given to Adam in the book of Genesis. Observing the receding of tide levels around Sweden he imagined this to be a worldwide feature that had allowed the original animals to multiply and migrate. He believed firmly in the fixity of species – the notion that all species had remained constant since the creation – commenting that

> there are as many species in existence as were brought forth by the Supreme Being in the beginning . . . and consequently there cannot be more species now than at the moment of creation.

However, when confronted with evidence of a plant which had deviated from one class into another, he acknowledged the possibility that new species could come into existence within the *plant* kingdom. But he made no concession to the notion of natural mutability between species of animals. His work of classifying and naming generally hardened the belief in the unchangingness of God-created life forms.[137]

Some biologists, however, regarded such classifications and divisions as rather artificial. They looked at the 17th century notion of a Great Chain of Being[138] and saw in it the minute differences between neighbouring elements rather than separate categories. They criticized too the morphological basis of his scheme – using bodily form as the guiding principle (the form of the sexual organs in the case of plants) – which made it seem bleak and static.

> To Buffon, and to everybody who reasoned like him, Linnaeus had killed living nature, whose real essence was dynamic force, change and everlasting movement in time.[139]

From the University of Uppsala the scene now changes to the Jardin du Roi in Paris where Linnaeus' contemporary critic and rival, Le Comte de Buffon, held sway.

GEORGES-LOUIS LECLERC, COMTE DE BUFFON (1707–88)

Buffon was a member of the Academy of Natural Sciences in Paris and the leading French naturalist of his day. He was not of noble birth but inherited a fortune which he then proceeded to increase greatly. In 1739 he was appointed director of the Jardin du Roi where he catalogued and built up a collection of plant and anatomical specimens. Between 1749 and 1788 he brought out his great work *Histoire Naturelle* in thirty-five volumes and after his death nine more volumes were added by a collaborator. It was one of the most impressive monuments of the French Enlightenment, highly influential on European thought during the second half of the 18th century, throughout the 19th century, and even beyond. The later part of that work included an imaginative account of the origin and development of the earth, *Epoques de la Nature* (1778).

Buffon shared the belief of de Maillet that the earth was formed over aeons by the slow action of natural causes, not by instantaneous divine action. Strangely, he seemed to think of the biblical flood as a universal supernatural event without any role in the unfolding of geological phenomena. Indeed, he criticised the 16th century English authors, Whiston, Burnet, and Woodward, for invoking the flood in their account of earth history, accusing them of distorting scriptural passages to fit geological theories and altogether producing a ridiculous mixture of natural history and theology. His defence of Scripture and acceptance of a supernatural status of the flood have surprised and puzzled historians as they have endeavoured to discover what he believed.[140]

Like many of his intellectual contemporaries Buffon was influenced by the heterodox theology of deism, seeking always to describe the natural world purely in terms of physical processes. As he put it:

> The Almighty gave a force of impulsion to the planets when he set the universe in motion; but we ought, as much as possible, to abstain in physics from having recourse to supernatural causes; and it appears that a probable reason may be given for this impulsive force, perfectly in accord with the laws of mechanics.

In his thinking about earth history Buffon, unlike Newton, sought a purely physical explanation for the common direction of rotation of the planets.[141] Buffon suggested that a comet had made an off-centre collision with the sun, knocking out a small part of its material as a group of fiery fragments which cooled down and became the planets we now observe. Testing experimentally the cooling rates of spheres of different materials and different sizes, he calculated how long it must have taken the planets to cool from white heat to a habitable temperature. This yielded an age of seventy-five thousand years for the earth, but from his geological knowledge he considered it to be much older, possibly as much as three million years.

Buffon divided his history of the earth into seven epochs in its long period of cooling. That it had cooled was surely confirmed by the remains of tropical animals in polar regions. It was only in the last of the epochs that humans made their appearance and they too, he believed, would eventually perish in a deep freeze. It was a world-picture constructed from a limited empirical base, seemingly in tune with the Genesis account of the successive days of creation.

On the biological front Buffon was an outspoken critic of Carl Linnaeus, especially over the latter's use of sexual anatomy as a criterion in his taxonomy, and over the seeming arbitrariness of the taxonomic hierarchy. The categories of classes, orders, and genera were simply human constructs, he argued, violating both the continuity of nature and the cardinal principle that nature only knows individuals.[142]

Buffon felt uncertain of the validity of both the *fixity* and the *mutability* of species. For example, he addressed the question whether the horse and ass had descended from a common stock, realising that if speciation (the transmutation of one species into another) had ever occurred, this would imply immense creative potential in the processes of nature. In the absence of intermediate forms,[143] his first conclusion was that such an occurrence was contrary to Scripture, reason and experience. Yet he wondered if gradual cumulative changes could take place, given a long enough period of time and the pressure of a changing environment.

It was clear that there is a temporal progression in the complexity of lifeforms, whether by special creation or natural process. Buffon avoided the dilemma by proposing that all life arose by spontaneous generation from 'organic molecules'. These had appeared through some chemical process or by the action of heat, formed themselves into assemblages and then given rise directly to higher forms like the elephant without any sort of evolution from simpler forms of life; so in this respect he was not a forerunner of Darwin. Nor, however, was he a supporter of the theory that all species had remained exactly the same since their appearance. For example he believed that the elephant had become smaller as the earth had cooled. As he suggested, fossil elephants (mammoths) were so large because the earth was then hotter and 'Nature in its first vigour' could generate such large creatures. Thus he related biological change to geological change. But he was not convinced that full scale transformation of species occurs, producing great variety from a few initial forms.[144]

Buffon holds an important place in the history of the development of concepts about evolution. He himself did not put forward any theory of his own nor, indeed, did he wholeheartedly support any such theory. His contribution lay rather in being the first to subject the idea of evolution to extensive criticism based on a wide range of empirical evidence.[145] He attempted a natural explanation of the origin of the earth in general, and of sedimentary rocks in particular, and thus did for Lamarck much what the eminent geologist Charles Lyell did for Darwin – he fostered the idea of a gradual temporal development due to natural causes.

In a review in the Jansenist journal[146] *Nouvelles Ecclesiastiques* during 1750, Buffon's *Histoire Naturelle* was portrayed as yet another attack upon Christianity. For example he had claimed, even if reluctantly, that one of the basic and humiliating truths that humans had to face was that they must be classed with the animals. Yet where was such a comparison to be found in the Bible? Did it not contradict the fundamental tenet that man is made in the image of God? Again, his theory of the formation of the earth and planets was nowhere to be found in the biblical account, yet he had the audacity to criticise other authors for a misuse of the Scriptures.[147]

But it was Buffon's ideas about the nature of truth that drew the strongest criticism. He had written that if mathematics is purely a construct of the human mind it follows that mathematical truths are somewhat ephemeral. He found physical science more firmly based, but seemed to regard moral truths as no more than matters of convenience. The reviewer was offended by such a cavalier approach to the question of truth, for the implications were dire: there would no longer be certainty about anything, no point in the adoration of God, no restraint from the practices of perjury, theft, adultery, and ultimately no protection against the destruction of religion and society.[148]

As a public figure Buffon felt he had to be careful what he wrote. He once confided that he feared theological controversy more than the criticism of scientists. When the same matters criticized in *Nouvelles Ecclesiastiques* were again singled out for censure by the Faculty of Theology at the University of Paris, Buffon responded immediately, agreeing to amend any paragraphs which the Faculty found to be contrary to the Faith. In the next volume of *Histoire Naturelle* he stated that he had no intention of contradicting the text of the Scriptures, that he firmly believed in the literal account of the creation in the Book of Genesis, and would therefore give up his theory of the formation of the solar system which, in any case, he regarded as no more than 'a purely philosophical supposition'. He also modified his remarks on truth, affirming that there were absolute and eternal truths, such as the existence of God, and the existence and immortality of the soul. A reply from the Faculty of Theology stated that his response had been received with extreme joy. However the *Nouvelles Ecclesiastiques* expressed scepticism about the genuineness of Buffon's recantation.[149]

Buffon moved within the milieu of *philosophe* scepticism. He had begun by accepting the hypothesis of the Great Chain of Being, but the notion had no greater hold upon his mind than any other piece of metaphysics; the traditions which influenced his thought were rather those of the ancient world of Greek thought, especially Epicurean atomism.[150] His religious beliefs seem elusive. He has been variously seen as an atheist, as a devout natural philosopher albeit a deist, and as a writer full of inconsistencies. A recent critic suggests that the frequent references to the Scriptures were intended simply to reassure the pious. At heart a deist, he no doubt shared the general anticlerical stance of the *philosophes* and their aversion to Church dogma.[151]

JEAN-BAPTISTE DE MONET, CHEVALIER DE LAMARCK (1744–1829)

The name of Lamarck is usually associated with a certain mechanism of evolution in the world of nature – the agency of *inheritance of acquired characteristics*. This was largely discounted by Darwin's notion of natural selection, yet Darwin and later biologists at times tended to invoke it as a subsidiary mechanism in biological evolution.

Lamarck was a French aristocrat by birth, but an impoverished one. As a young man he served in the army, was invalided out, and turned to the study of medicine and biology. Under the patronage of Buffon he acquired a position at the Jardin du Roi in Paris in 1788, and was appointed six years later to a professorship at the great Museum of Natural History. He has been described as restrained, perhaps a trifle prim, extremely conscientious yet failing to realise the extent to which his theorizing lacked experimental support. Indeed, it is surprising that so competent and observant a naturalist never carried out an experiment. As one writer has put it, his theory of evolution, however influential for a while, was unhampered by any too nice an attention to evidence.[152]

Lamarck brought to his study of natural history a deistic view of harmony in the universe – harmony arising from underlying design. His writings show, too, the continuing influence of the idea of the Great Chain of Being. In an early publication, *Hydrogeologie* (1802), he refers casually to an earth history stretching over millions of centuries, an assumption favourable for the evolution of species. Time, he wrote, was never a difficulty for nature. He thought of the Chain of Being as an 'escalator of being'. At the foot of the moving staircase simple unicellular organisms were fed in through spontaneous generation from non-living matter. These gradually rose by an evolutionary process of nature over aeons of time to become human beings at the top of the escalator.

An early version of his theory of evolution appeared in 1801 in which he seems to have attributed the evolutionary process entirely to the agency of the inheritance of acquired characteristics, often called the Lamarckian mechanism. Here he suggested that the structure of any species of animal changes over time in response to changes in the environment or as the result of the spontaneous formation of new habits. For example, the need to pass through confined spaces has elongated the snake's body, and the giraffe has developed a long neck by constantly straining upwards to eat from the trees. Furthermore, any such changes during the creature's lifetime are passed on to its offspring. From genetics we know now that this is not the case – heredity is governed primarily by the passing on of genes, not acquired characteristics – but at the time it seemed a reasonable hypothesis.[153]

Lamarck's best known book *Philosophie Zoologique* (1809) was the first full statement of his evolutionary views, aiming to show how biology (a term he helped to popularize) could not only classify animals but also explain a great deal about them. Here the primary principle in nature is its inherent tendency to produce organisms of increased complexity.[154] In Lamarck's view this law of nature was driven by some inner force which operated continuously to improve the species. His inheritance mechanism now took second place, acting sometimes to disrupt the complexification trend and to produce side branches to the main chain of animal forms.

In 1815 Lamarck published his last major treatise, *The Natural History of Invertebrate Animals*.[155] Here, although still influenced by the old Chain of Being, he rejected the idea of a continuous gradation of forms from nonliving to living things. Within the chain of living things, how did transformation occur from one species to the next on the scale?

He suggested the operation of four laws:

- Nature tends to increase the size of living individuals to a predetermined limit.[156]
- The production of a new organ results from a new need.
- The development reached by organs is proportional to the extent to which they are used.
- Everything acquired by the individual is transmitted to its offspring.

Lamarck continued to see the whole process as entirely mechanistic. Changes in the environment created new needs and therefore new habits. These led to the stimulation and movement of interior fluids which, in turn, acted to modify the bodily structure. Underlying all the processes of nature, however, was the creative intention and power of the Creator. He expresses it thus:

> The general power which holds in its domain all the things we can perceive is truly a limited power, and in a manner blind; a power

> which has neither intention, nor end in view, nor choice; a power which, as great as it may be, can do nothing but what in fact it does; in a word a power which only exists by the will of a higher and limitless power, which, having founded it, is in truth the author of all that it produces, that is, of all that exists .... And *nature* is only an instrument, only the particular means which it has pleased the *supreme power* to employ in the production of the various bodies, in their diversification; to give them properties, or even abilities .... She is, in a way, only an intermediary between God and the parts of the physical universe, for the execution of the divine will.[157]

Commenting on the common structure of the vertebrates he writes:

> Passing from the fishes up to the mammals, we see that this plan becomes more perfect from class to class, and that it only reaches its final form in the most perfect mammals; but we also observe that, in the course of reaching its perfection, this plan has suffered numerous modifications (from a regular and fine gradation) ... on account of the places in which the animals live and the habits which each race has been forced to acquire according to its circumstances.[158]

Altogether he maintained the notion of a neatly planned universe. In his later writing he attempted to set up a linear array of animal forms, representing a rising scale of perfection. But he rejected forcefully the notion of a single chain linking all the bodies produced by nature. Animals and plants constitute two independent lines, their differences being due to differences in the chemical make-up of the primordial plant and primordial animal respectively. He attributed the sharp break between the living and the non-living to a difference in level of organization.

Lamarck was dubbed by the English 'the French atheist' and his eager support of the French Revolution only confirmed their distaste. They preferred to put their faith in the intervening hand of God who had created the highest animals as directly as the lowest; there was no changing of one into the other.[159]

GEORGES CUVIER, BARON (1769-1832)

The work of Georges Cuvier on the rock strata of the Paris Basin and his invention and skilful application of the new science of comparative anatomy to the fossil remains of prehistoric monsters made him the leading naturalist of his day. He was highly influential in French intellectual life and in later years played the role of an official parliamentary spokesman.

Cuvier overshadowed Lamarck, at times clashing with him vehemently on the topic of evolution. He rejected the hypothesis of the transmutation of species, not so much on religious grounds as from his reasoning as an anatomist. He had acquired a deep grasp of the complexities of living organisms and the way their conditions of existence impose necessary links between the internal structure of bodily forms and the external environment. Indeed, it was said of him that he could in imagination reconstruct the whole animal from a single bone. Given the delicate interdependences within each animal, he believed that any significant change in these would make its life unviable. Even when reconstructing extinct forms from their fossil remains, he insisted that these remained as distinct species that could not be tranformed into new ones. Cuvier thus presented the fixity of species as a pragmatic consequence of his concern for the complexity of living things, and his writings only rarely expressed his faith in a supernatural Designer.[160]

In introducing the idea of placing together all vertebrates (creatures with backbones) as a basic division of the animal realm, Cuvier introduced a new sense of the quite different ways in which vertebrates and invertebrates live and function. This in turn tended to break down the idea of linear ranking inherited from the old Chain of Being that still influenced Lamarck. Peter Bowler explains:

> It was many decades before the majority of naturalists could bring themselves to accept this complete breakdown of the old hierarchical viewpoint, but the implications for the rise of evolution theory were enormous. It no longer would be possible to think in terms of a linear progress through the animal kingdom; each group

> would have to be pictured as a separate branch in a treelike process of development.

Having argued for the fixity of species, Cuvier wondered why it was that so many became extinct. He found an answer in his exploration of the Paris basin, in which the stratification of fresh and salt water deposits indicated that there had occurred a series of major changes in the relative positions of sea and land. These appeared to have been comparatively sudden because there were abrupt breaks between fossil populations. The extinctions must have resulted from catastrophic geological changes, he believed, wrought by far more powerful forces of nature than we now observe. Thus his researches convinced him that the principle of uniformity was not tenable. He then summarized the history of the earth in the following passage:

> The lands once laid dry have been reinundated several times, whether by invasions of the sea or by transient floods; and it is further apparent to whoever studies the regions liberated by the water in its last retreat, that these areas, now inhabited by men and land animals, had already been above the surface at least once – possibly several times – and that they had formerly sustained quadrupeds, birds, plants, and terrestrial productions of all types. The sea, therefore, has now departed from lands which it had previously invaded.
>
> ... It is also extremely important to notice that these repeated inroads and retreats were by no means gradual. On the contrary, the majority of the cataclysms that produced them were sudden. This is particularly easy to demonstrate for the last one which, by a double movement, first engulfed and then exposed our present continents. ... It also left in northern countries the bodies of great quadrupeds, encased in ice and preserved with their skin, hair, and flesh down to our own times. If they had not been frozen as soon as killed, putrefaction would have decomposed the carcasses. And, on the other hand, this continual frost did not previously occupy the places where the animals were seized by the ice, for they could not have existed in such a temperature. ... Life in those times was often disturbed by these frightful events. Numberless living things were victims of such catastrophes: some, inhabitants of the dry land,

> were engulfed in deluges; others, living in the heart of the seas, were left stranded when the ocean floor was suddenly raised up again; and whole races were destroyed forever, leaving only a few relics which the naturalist can scarcely recognize.[161]

Elsewhere Cuvier estimates that the last great revolution in the earth's surface structure occurred five or six thousand years ago (calculating this from the rate of increase of deltas, sand dunes and peat deposits)[162] – and that the small number of individuals of men and other animals that escaped its effects have since propagated and spread over the lands then newly laid dry; and, consequently, that the human race has only resumed a progressive state of improvement since that epoch. He refused to support those who identified this dramatic event with the universal deluge of the Bible, for he regarded the changes in the boundaries of land and sea as the result of strictly localized catastrophes. He believed that at some earlier time all extinct and living animals had coexisted around the world and that some were wiped out by successive catastrophic events. However, fossilized remains of *modern* animals were nowhere to be found, so geologists concluded that new forms of life had emerged on earth at many different moments in its history.

Cuvier's writings prepared the way for the school of geological thought that would later be named *catastrophism*.

JAMES HUTTON (1726-97)

The *uniformitarian* approach to earth history is well exemplified by James Hutton (1726-97), who has been referred to as both the founder of modern geology and perhaps the most typical of the deistic geologists. He and his close friend Joseph Black were by far the most capable Scottish scientists of their day, who both happened to be non-practising doctors of medicine. No doubt Black's expertise in the physics of heat contributed to Hutton's notion of heat as a geological (volcanic) agent, alongside the processes of erosion and precipitation by wind and water.

In his book *Theory of the Earth* (1795) Hutton presented a picture of the earth itself as a stable system, undergoing slow cyclical changes through the combined effects of high temperature and high pressure – always by the presently observable processes of nature rather than divinely imposed cataclysms. He argued against the catastrophist assumption of the occurrence of largescale deluges, since the purpose of the earth is evidently to maintain vegetable and animal life, not to destroy them. With his uniformitarian assumption of unimaginably long periods of geological formation he came to the controversial conclusion that 'we find no vestige of a beginning, no prospect of an end'.[163] This brought charges of atheism for if, indeed, there were no beginning to the world, what place did that leave for a creator? But he was a deist rather than an atheist.

Hutton's deism found expression in the idea of the earth being evidently made for man. In fact he was more concerned to obtain knowledge of final causes than to explain physical phenomena. His main aim was to discover *order* in nature – 'order not unworthy of Divine wisdom'. On one hand he rejected the notion of natural processes as the outcome of chance, and on the other he was equally opposed to catastrophism, for any cataclysmic destruction seemed an unnecessary element in the earth's development. Moreover his deism freed him from the constraint of 'biblical geology', allowing him to contemplate a very great age for the earth. Conversely his geology helped him to see the presence and efficacy of design and intelligence in the power that conducts the world, that is, the 'Author of Nature'.

John Playfair, an Edinburgh geologist and mathematician, supported Hutton's emphasis on order in nature for, as a minister in the Church of Scotland, he saw this as showing greater respect for the Creator's work than the misconceived effort to demonstrate the geological credibility of every word of Genesis. Playfair was critical of Deluc and Kirwan 'who would have us consider their geological speculations as a commentary on the text of Moses'. This was as injurious to the dignity of religion as to the freedom of philosophical inquiry, whereas Hutton's emphasis on order in nature, with its rythm of geological processes, forever building new substances and new worlds out of old and worn-out materials, constituted a useful contribution to natural theology.[164]

Hutton's work elicited a robust rebuttal from Richard Kirwan, who was president of the Royal Irish Academy from 1799 to 1819 and has been described as the father of British mineralogy. In his *Geological Essays* (1799) Kirwan claimed that the function of geology is to lead one into religion, as religion leads one into morality. In his view geology provided an underpinning to Mosaic history[165] which had hitherto been missing. He claimed that this absence had proved 'favourable to the structure of various systems of atheism or infidelity, as these have in their turn to turbulence and immorality'. In the previous century Boyle too had considered that a cavalier exegesis of the Book of Genesis could open the way to a disastrous escalation of atheism and then immorality. For Kirwan the urgent task was to deal with this particular ignorance, 'to dispel it by all the lights by which modern geological researches have struck out'.[166]

In Hutton's day a school of Neptunist geology had been formed recently by Abraham Werner, a German professor of mineralogy. Werner's main idea was that the earth was at one stage completely covered by water in which were suspended or dissolved the component materials of the earth's surface. He postulated five stages of deposition of this material to form successive strata. It was a detailed and beautifully simple explanation but it presupposed the movement of water on an immense scale and could not account for certain key observations. For example, the order of mineral contents in rocks differed from place to place, and some rock formations showed older strata lying on top of more recent ones (as determined by Werner's age measurements and by others that were made later). As a purely scientific hypothesis (not linked to the biblical flood), the Wernerian scheme faded out in the early 1800s.

WILLIAM BUCKLAND (1784–1856)

Cuvier's ideas appealed to William Buckland and his fellow catastrophists. Buckland was England's best known naturalist and popularizer of natural science in the 1820s, having been appointed as the professor of geology at Oxford in 1819. He would have endorsed the sentiment of his opposite number at Cambridge, Adam Sedgwick, that geology, like every other science when well interpreted, helps to promote 'natural religion'. He was not only an enthusiastic lecturer (wont to meet his classes at the sites of their field work wearing top hat and academic gown) but also a serious apologist for the claims of natural religion. In his inaugural lecture (*Vindicae Geologicae; or, the Connexion of Geology with Religion Explained*) he aimed to show that the study of geology has a tendency to confirm the evidences of natural religion and that the facts developed by it are consistent with the accounts of the creation and deluge recorded in the Mosaic writings.[167]

At a time when William Paley's design arguments (for the existence of a beneficent Creator and Sustainer of the universe) were still widely enjoyed, Buckland devoted part of his inaugural lecture to the interdependence of flora and fauna and to the nicely arranged position of the earth in the solar system. 'He must be blind indeed,' he asserted, 'who refuses to recognise in them proofs of the most exalted attributes of the Creator'. At one point he went so far as to speak of the happy relations between rock structure and mineral resources as an illustration of the way the 'Omnipotent Architect' or 'Divine Engineer' had assured primacy to British manufacturers.

For the next four decades there was a ready market for such writing by various fellows of the Royal Society, vying to illustrate 'the web which knitted the existence of a God to the unity of design exhibited in His works'. Indeed, it seemed that this had become a way of looking not only at the universe but also at society, seeing in every detail of natural *and* social relations the evidence and hence the sanction of divinely ordained natural necessity.[168]

For Buckland, the Creator was *not* the deistic God of Hutton (and Voltaire and Erasmus Darwin, grandfather of Charles). On the contrary, geology was pointing to the God who intervened in the history of Israel, as well as in nature; both interventions were considered miraculous.[169] Consequently, Buckland, Sedgwick and other catastrophists are often regarded as standing in opposition to Hutton and his successors, in terms not only of geological theory but also of their underlying theological assumptions. It may be appropriate, though, to class them as *semi-deists* for they divided the world into two compartments: a virtually deistic part in which physical law reigned supreme, and an 'interventionist' part which was the sphere of action of the God of theism.[170]

What bothered Buckland was the often expressed scepticism of deists concerning any continuing divine action in the world. They claimed that the universe is carried on by the force of the laws originally impressed on matter, without the necessity of fresh interference or continued supervision on the part of the Creator. Such interference, claimed Buckland, was surely to be seen in the geological convulsions that have operated at successive periods, not blindly and at random, but 'with a direction to beneficial ends'. Thus does an overruling Intelligence continue to 'superintend, direct, modify and control the operations of the agents which he originally ordained'.[171] Buckland had taken over Cuvier's strictly scientific catastrophism and re-interpreted the catastrophes and the sudden new faunas as signs of the immediate exercise of divine power.

These ideas were already in place when they were confirmed, it seemed, by the dramatic discovery in 1821 of the Kirkdale Cave in Yorkshire, a treasure trove of fossil bones embedded in mud. Buckland carefully and skilfully established that this had been a den of hyenas, whose bones lay among those of their victims – lions, tigers, elephants and others, all long extinct in Britain.[172] The presence of mud and absence of such fossils except in secluded caves amounted to strong evidence for an immense flood – as described in Genesis – which had swept away almost all those early fauna. Within two years Buckland and his colleagues had explored more than twenty caves in England and the Continent, finding similar fossil bones buried in diluvial silt. The outcome was a scientific paper, honoured by the Royal Society, and a book *Reliquiae Diluvianae; or, Observations on the Organic Remains Contained in Caves, Fissures and Diluvial Gravel, and on Other Geological Phenomena, Attesting the Action of an Universal Deluge* (1823), which met with resounding praise.[173]

ADAM SEDGWICK (1785–1873)

Within a decade or so, Buckland, Sedgwick and the geological community as a whole had abandoned this explanation of the Kirkdale

Cave finds. Counter evidence had been accumulating to suggest that it was glacial action which had caused that extinction. In 1825 Sedgwick had asserted that a few thousand years ago the earth's surface was submerged by the waters of a general deluge. The investigations of geology, he claimed, tend to prove that the accumulations of alluvial matter have not been going on for many thousands of years, and that they were preceded by a great catastrophe which left traces of its operation in the *diluvial detritus* which is spread out over all the strata of the earth. Six years later, however, in a final statement as President of the Geological Society, Sedgwick explained that he had changed his mind completely:

> Having been myself a believer, and, to the best of my power, a propagator of what I now regard as a philosophic heresy ... I think it right, as one of my last acts before I quit this Chair, thus publicly to read my recantation. We ought, indeed, to have paused before we first adopted the diluvian theory, and referred all our old superficial gravel to the action of the Mosaic Flood. For of man, and the works of his hands, we have not yet found a single trace among the remnants of a former world entombed in these deposits.

It should be noted that this turn-around was confined to the purely scientific aspects of the debate. For Sedgwick there was no doubting the historicity of Noah's flood, for he remained committed to what he termed evangelical biblical doctrines, looking to Scripture as providing an authoritative account of divine will and action in history.[174] He was also deeply committed to geology as a science of *observation*, raising objections against the adoption *a priori* of Hutton's principle of uniformity and the consequent idea of endless repetitive cycles in the history of the earth.

CHARLES LYELL (1797-1875)

Charles Lyell was a London barrister-turned-geologist who held the chair of geology at King's College London during the years 1831-33 and then resigned because he enjoyed neither the prestige nor the remuneration he had expected from the post. In that period he was much engrossed in writing his seminal three-volume work, *Principles of Geology, being an Attempt to Explain the Former Changes of the Earth's Surface by Reference to Causes now in Operation.* The third volume included the latest developments in stratigraphy, paleontology and physical geography, and included his important contribution to the establishing of earth history in his identification of the Pliocene, Miocene and Eocene epochs of the Tertiary period (which ranged from about 60 million to 5 million years ago).

His ideas seem to have crystallized fairly rapidly in the late 1820s, as he drew upon recent descriptions of volcanic formations and toured the Continent extensively, making a close study of Mount Etna (Sicily). Here he found the volcano structured in layers which had accumulated to a height of over three thousand metres, lying upon relatively young strata. Furthermore, similar strata were to be found at the centre of Sicily at an elevation of nearly a thousand metres. Reflecting upon these data, together with recent speculation that volcanic and other geological processes had not been more violent in earlier phases of earth history, he decided on an extreme uniformitarian framework for his theorizing. He assumed not only that the agents of geological change have always been those now observed, but also that they have acted throughout with the same intensity. He then sought to construct the history of the earth by identifying the successive (though not necessarily progressive) states through which it had passed and the causal effects involved, bringing to bear upon his interpretation a greatly expanded repertoire of geological forces.[175]

In the second volume of his *Principles of Geology*, Lyell went on to suggest that life on earth had always existed in some form or other. Here he invoked the occurrence of certain marine life forms throughout the fossil record as an indication that the conditions for such forms kept recurring. Noting, too, that no mammalian fossils had ever been found in the earliest rock strata, he argued that absence of evidence could not be taken to imply non-existence; perhaps the chances of their being preserved were always slim. He sought throughout to strip the fossil record of any progressive pattern.[176]

Lyell's passionate upholding of the uniformity of nature throughout time thus extended to both the inorganic and organic realms of nature. He believed that each species maintained its distinct identity, despite minor variations within it. The populations of a given species might fluctuate and even disappear from time to time, starting afresh at different times and places through the creation of a new male-female pair. He seemed to allow the idea of divine intervention at this level, even if not in the detailed forming of the earth's surface.

His main concern was to establish geology as a rigorous science, free from the distorting effect of any theological presuppositions or speculation about origins, purposes and meanings. Consequently he has often been regarded as a secular hero of science who rescued geology from biblically-oriented catastrophism, especially from the notion of the biblical flood as an historical event.[177]

Lyell's strong uniformitarianism certainly aroused severely critical comment from Adam Sedgwick, his colleague in the Geological Society of London. Sedgwick pointed out that volcanic forces, in particular, were of great complexity and almost impossible to calculate; therefore the denial of any cataclysmic intensity in the past seemed a merely gratuitous hypothesis.[178] But what Sedgwick and his botanist colleague John Henslow critcised most in Lyell's writings was the basic theme that there is no evident direction, progression or purpose in the history of nature. To eliminate the role of the Genesis flood from the geological account of earth history was regrettable, but to eliminate the notion of continuing divine purpose and action in the world was worse, for it opened the way to moral and social decay. After attacking this idea in his final presidential address to the Geological Society, Sedgwick concluded that 'Mr Lyell has, unconsciously, been sometimes warped by his hypothesis'.[179]

Lyell's stand against catastrophism has been interpreted as that of a crypto-Unitarian deist against the overt 'biblical geology' of Buckland and Sedgwick – of the liberal thinker against Anglican orthodoxy. However, a deeper explanation may lie in the distaste he felt for *progressive transformism* when, in 1827, he read Lamarck's theory that species slowly but steadily improve in their adaptation to their environment, each generation passing on its acquired improvements to the next, eventually changing into a new species. Perhaps it was this hypothesis that triggered his opposition to all ideas of progress and drew him to a steady-state model of nature in which each species continually appeared and disappeared. What he disliked about Lamarck's doctrine of the inheritance of acquired characteristics was that it gave room to the picture of man as simply a glorified orangutan – which he regarded as an affront to human dignity. Lyell attacked the doctrine of Lamarck, using arguments largely modelled on those of Cuvier.[180] Nevertheless, his dislike of the progressive development of species should not blind us to the fact that his own sustained account of the effect of geological forces on the distribution, dispersal and extinction of species prepared the ground more thoroughly than any predecessor for the kind of questions that Darwin and Wallace were to ask.[181]

It is not immediately evident what influence Lyell's beliefs may have had on the development of his scientific views. If his religion was essentially a-historical, would this have encouraged the a-historical motifs of his geology?[182] A high view of humankind may have encouraged a steady-state model of biological species; on the other hand, theological unorthodoxy may have freed him to ignore the biblical flood.

There has been a tendency in English accounts of the history of geology to concentrate on Lyell as the great unifier of 19th century theories and arguments in this field. Recent writers unfold a science deep in controversy about issues of fact, interpretation, method and even aims. The geologists of continental Europe worked rather independently of their English counterparts and produced a comprehensive theory of earth history which suggested the occurrence of catastrophic periods of disturbance separated by periods of quiet readjustment. This relied on several strands of evidence that the surface layer structure of the earth on both sides of the Atlantic is the same – the result, it appeared, of the earth's nebula origin and the subsequent cooling that progressively destroyed the continental crust and pointed to a future static earth covered by a universal ocean.[183] This theory grew to worldwide acceptance by 1900 and then rapidly declined and collapsed, soon to be replaced by a tentative theory of plate tectonics, namely that the earth's crust comprises several large plates capable of slow relative movement, observed as continental drift.[184]

If Lyell can no longer be claimed to be the Newton of geology, he certainly played a vital role in encouraging the work of Charles Darwin, who not only drew heavily upon Lyell's *Principles of Geology* during his epic voyage in HMS 'Beagle' but also relied on him as a mentor through the development and launching of his theory of evolution.

CHARLES DARWIN (1809-1882)

With the development of natural history during the early 19th century, the question of the compatibility of new scientific understandings with the biblical account continued to arise. Did one look for a 'biblical geology' and a 'biblical zoology', or was all scientific theorising to be allowed to become free of biblical constraints? Just as geologists Buckland and Sedgwick were influenced by the Genesis account of the flood, so too were the biologists of that period mindful of the creation narrative: 'And God made wild animals, cattle and every creeping thing, all according to their various kinds'. Despite Buffon's influential doubts, the *fixity of species* seemed strongly confirmed by the taxonomic work of John Ray in the 17th century, Carl Linnaeus in the 18th, and Georges Cuvier in the early 19th.

The notion of fixity was contradicted by Darwin in his theory about the evolution of species by means of natural selection – a theory of how species can be transformed into new ones, not how they originated. Not only did this stir up heated debate at the level of its scientific merits but it was both supported and opposed on theological grounds and, indeed, seemed to lie at the heart of a change of attitude that has been described as one of those rare shifts that have occurred only half a dozen times in the whole of intellectual history.[185]

Charles Darwin (1809-1882), son and grandson of notable doctors, began to study medicine at Edinburgh University but changed to the general degree course at Cambridge. At that stage he held to orthodox religious beliefs and intended to be ordained in the Anglican Church. He regarded his academic studies largely a waste of time but was impressed by Paley's *Natural Theology*. He continued to develop his long-standing interest in the world of nature and attended lectures by the new young professor of botany, John Henslow. The two became close friends and in due course

Henslow recommended the 22-year-old to Robert Fitzroy, captain of HMS 'Beagle', for the role of captain's companion and natural historian on an intended two-year voyage of exploration along the South American coasts. This turned out to be a round-the-world epic that Darwin described as by far the greatest event of his life, lasting from December 1831 to October 1836 and setting him on an illustrious illness-plagued career.

Darwin had been well introduced to the discipline of geology at Cambridge through his informal connection with Professor Adam Sedgwick, and his main reading matter on the voyage was Lyell's *Principles of Geology*.

> Lyell's glittering prose style, his rich and varied knowledge, his succinct synopses of what was already known about the structure of the earth, his tart characterisation of current controversies, ... his inclination to discuss large issues and ask fundamental questions without risking the label of atheist or political radical, the clarity of his vision, his wit, and the well integrated nature of the geological history that he evoked were avidly absorbed.[186]

Indeed, the book was crucial to Darwin's thinking over a broad sweep of natural history and it has been claimed that without Lyell there would have been no Darwin, no intellectual journey, and no voyage of the 'Beagle' as commonly understood. His influence on the young traveller can hardly be overestimated.[187]

When the 'Beagle' reached the island of St Jago in the Cape Verdes, with its abundant tropical vegetation and remarkable geological features, he was able to bring Lyell's ideas to bear on the scene and produce a plausible account of the island's geological development. Later, during the South American part of the voyage, Darwin was to devise a theory of the formation of coral reefs based on his understanding (from Lyell) that the elevation of land, such as the Andes, was bound to be balanced by comparable subsidence elsewhere, the geological ups and downs together constituting one of Lyell's self-regulating oscillatory movements of the earth's crust.[188] On his return to England six years later, as a naturalist whose exploits were already becoming known among English scientists, it was in the Geological Society of London that he found his first scientific niche, as its general secretary.

In South America he travelled extensively, saw something of the horrors of slavery and the maltreatment of native peoples by their Spanish and British masters,[189] carried out extensive geological fieldwork and collected specimens of many kinds, including fossil bones of large extinct animals peculiar to the continent – giant armadillos, wild llamas and gigantic sloths, for example. He was fascinated by his glimpse of the primitive people of Tierra del Fuego.

> I shall never forget how savage and wild one group was ... absolutely naked and with long streaming hair ... they sent forth most hideous yells. Their appearance was so strange that it was scarcely like that of earthly inhabitants.[190]

No doubt in later years he pondered the world of difference – a world of improvement – between, as he put it, the faculties of a Fuegian savage and a Sir Isaac Newton. Then and throughout the voyage he kept detailed field-notes and a diary, from which he soon wrote a much acclaimed journal, describing his researches in both geology and natural history.

It was the visit to the Galapagos Islands, a year before the end of the voyage, that played a crucial role in focusing his thoughts on the big question of the evolution of life-forms. These islands were of recent volcanic origin, all similar in climate, situated on the equator a thousand kilometres to the west of South America. They were known to possess a rich variety of species that lived nowhere else in the world. Central to his subsequent thinking was the observation that there existed *slight variations within a given species*, especially in the populations of birds and giant tortoises that he studied there. He noticed that the finches, for example, showed differences in structure or colour, yet all belonged to the same species. It seemed too that each species had developed slightly differently on each island – but he kept wondering how this had happened.

In the second volume of his *Principles of Geology*, Lyell had speculated on the origins of island species, within a general assessment of the patterns of distribution of species over the planet. He was strongly biased against any notion of the transmutation of species, launching a point by point onslaught against Lamarck's scheme. He considered such ideas not only impossible in practical terms but also theologically subversive, for he was anxious to safeguard the special status of humankind. Despite his concern that geology could only become a science when it became independent of biblical ideas, Lyell believed in the 'special creation' of human beings and was anxious to dismiss any idea of a close link between humans and the animal world. Nevertheless, his acute summarizing of the outstanding problems in biology was to act as a useful foil for Darwin's nascent ideas on transmutation.

Within a few months of his return to England Darwin opened his first note-book on the Transmutation of Species and, with sufficient inherited wealth to free himself from the need to earn a living, began his long assault on the fixity of species. During 1838 he encountered Thomas Malthus's *Essay on the Principle of Population* (1798) which provided him with a crucial idea for his slowly forming theory. Malthus proposed that the rate of population increase, if unchecked, will always outstrip the rate of food production. Consequently there is a struggle for existence, not only among human beings but throughout the world of nature. In almost all situations there are many offspring that do not survive to the stage of producing their own offspring.

Darwin learnt a good deal about the breeding of new varieties of pigeons, dogs and horses – the deliberate selecting from the available range of variations in order to enhance certain features over successive generations. But what was it, he wondered, that played the role of breeder in the wild, enabling varieties to change into new varieties – and perhaps even into new species? On reading Malthus, he wrote:

> (I)t at once struck me that under these circumstances (in the struggle for existence) favourable variations would tend to be preserved, and unfavourable ones to be destroyed. The result of this would be the formation of new species. Here, then, I had at last got a theory by which to work.[191]

As environments changed or as food became scarce, species would gradually change as, on average, the better adapted survived and the less well-adapted did not. Here was a plausible mechanism, which he called 'natural selection' (in contradistinction to the 'artificial selection' performed by breeders), which enabled him to begin to theorize over a wide range of biological phenomena, even if the contemporary understanding of heredity that he adopted was incorrect.

The outline of Darwin's theory was in place by 1842, and within two more years he had produced a 230-page manuscript which he believed was sufficiently developed to be published in case of his early death. He worried about how the public would react to his revolutionary ideas and about the feelings of his father and devout wife. He wanted time to test his ideas against various kinds of evidence and, often in poor health, he preferred the ongoing enquiry to the task of completing the book on Natural Selection that was begun in 1856 but never finished.[192] Eventually he was jolted into allowing his theory to be made public when a younger naturalist, Arthur Russel Wallace, hit upon the idea of natural selection and, unaware of Darwin's intellectual journey, sent him a manuscript that was uncannily similar to Darwin's 1842 outline. Through Lyell and botanist Joseph Hooker, Darwin's and Wallace's ideas of natural selection were presented jointly – in their absence – at a meeting of the Linnaean Society in July 1858.

Almost immediately Darwin set about writing *The Origin of Species*.[193] After an intensive effort amidst much groaning of spirit, sickness, impatience and anxiety, he completed the book in thirteen months and it was published in November 1859. He was anxious about the response of the public for, without overtly stating so, the book raised fundamental questions about humankind and our place in the natural world. But what he wanted most of all was that the book should be judged on its scientific merits. The first edition (1250 copies) was entirely subscribed for on the day of publication and there were to be five more editions over the following thirteen years, with considerable revisions, amplifications, corrections and new material, some of it in response to the questions and comments of fellow scientists.

In 1871 Darwin's further major work, *The Descent of Man, and Selection in Relation to Sex*, was published in two volumes and 4500 copies were sold within the first couple of months. What was clearly implicit in *The Origin of Species* was now made explicit: humankind's intimate link to the natural world, subject to natural selection like any other species. He wrote:

> Not only does man share the same physical lay-out and processes, but mentally man and the lower animals do not differ in kind, although immensely in degree. ... man is constructed on the same general type or model as other mammals. All the bones in his skeleton can be compared with corresponding bones in a monkey, bat, or seal. So it is with his muscles, nerves, blood-vessels and internal viscera. The brain, the most important of all the organs, follows the same law, as shewn by Huxley and other anatomists.

Further similarities were evident, such as the intercommunicability of diseases between humans and other mammals, the sameness in the processes of reproduction and embryonic development, and the presence of a vestigial tail (the coccyx) in the human body.[194]

Darwin had already considered sexual selection as a fundamental factor in the development of species, alongside natural selection – the choosing of mates, whether by the males or the females. Two-sex systems have the advantage (over hermaphroditism or the fertilization-less process of parthenogenesis) of producing the greatest degree of variation in offspring and therefore lend themselves to the processes of natural selection. A striking thing about sexual selection is its aesthetic aspect: the appreciation of beauty by human and non-human animals alike. The topic of sexual selection occupies the entire second volume of *The Descent of Man*, emphasising Darwin's larger argument that even man's most unique attributes find analogies among the beasts.[195]

Although Darwin's account of nature's evolutionary process was seized upon and utilized in different ways by a variety of groups – Fascists and Marxists, imperialists and anarchists – he himself insisted always on limiting the extra-scientific implications of his work and resisted any overt politicization.[196] This was especially true of the theological implications and he refused to be drawn into advocacy of atheism, preferring to apply to himself the word *agnostic*, coined by Thomas Huxley. In correspondence with the prominent Presbyterian botanist at Harvard University, Asa Gray, just after the publication of *The Origin*, he commented:

> I had no intention to write atheistically. But I own that I cannot see as plainly as others do . . . evidence of design and beneficence on all sides of us. There seems to be so much misery in the world. I cannot persuade myself that a beneficent and omnipotent God would have designedly created the *ichneumonidae* with the express intention of their feeding within the living bodies of caterpillars, or that a cat should play with mice . . . . On the other hand I cannot anyhow be contented to view this wonderful universe, and especially the nature of man, and to conclude that everything is the result of brute force. I am inclined to look at everything as resulting from designed laws, with the details, whether good or bad, left to the working out of what we may call chance . . . . But the more I think the more bewildered I become.

What had become of the orthodox religious beliefs that Darwin brought to Cambridge – beliefs that he seemed to carry through to the end of his five-year voyage on *The Beagle*? Writing an autobiographical piece for his family forty years later, he explained that just as the idea of a clerical career died a slow natural death, so his belief in Christianity as a divine revelation withered gradually, deeply shaken by the death (in 1851) of his young daughter Annie. He drifted from orthodoxy to liberal beliefs in God, thence to an agnosticism which at times turned into atheism. He could not see how anyone ought to wish Christianity to be true.

> If it were, the plain language (of the New Testament) seems to show that the men who do not believe, and this would include my Father, Brother and almost all my best friends, will be everlastingly punished. And this is a damnable doctrine.[197]

As it was, many Victorian intellectuals struggled with the problem of Christian belief, and many fiercely rejected Evangelical or Catholic orthodoxy in its scheme of salvation, especially notions of divine favouritism, the substitutionary atonement, and everlasting torment in hell. Yet near the end of his life Darwin could criticise a prominent atheist for declaring his beliefs aggressively. Is anything to be gained, he asked, by forcing new ideas on people, especially ordinary people who may not be ripe for it?[199] Besides, he was utterly devoted to his wife Emma whose deeply held Christian belief he respected. Ultimately he seemed to draw back from complete unbelief, sensing 'the extreme difficulty or rather impossibility of conceiving this immense and wonderful universe ... as the result of blind chance or necessity'.

In spite of continuing bouts of poor health, Darwin lived out another decade in his country home in Kent, enjoying being with his family and working on further minor biological projects. When he died in April 1882, there was a remarkable outpouring of tributes from far and wide and the science community in London quickly arranged, with the family's eventual consent, that Darwin be buried in Westminster Abbey, beneath the monument to Newton. Was he not 'the greatest Englishman since Newton'? Had he not given 'exactly the same stir, the same direction, to all that is most characteristic in the intellectual energy of the nineteenth century, as did Locke and Newton in the eighteenth?'[200]

DARWIN'S THEORY OF EVOLUTION

Darwin's theory was based on three reasonable assumptions:

- *hyperproductivity* (or *super-fecundity*): organisms produce more offspring than can reach maturity;
- *variability*: a range of differences exists within any species (in anatomical details which can determine, for example, the ability to see, move, digest, hide from predators, etc); and
- *natural selection*: environmental change, food shortages, and the presence of predators together create for a species a struggle for existence, gradually and inexorably weeding out the less well adapted members through successive generations, while allowing the better adapted to survive to the stage of procreation, and thus pass on their favourable characteristics.[201]

It was the combination of *variability* and *natural selection* that was Darwin's key insight. The many slight differences within a species are highly important, he explained, as they afford materials for natural selection to work on, just as the breeder of domestic animals or birds can accumulate individual differences in any given direction by artificial selection. Without such slight variations natural selection could make no headway. Ultimately the operation of natural selection makes for the adaptation of organisms to their environment, giving the appearance of superb design.

The heart of Darwin's theory, natural selection, is presented in the fourth chapter of *The Origin of Species*. He begins the chapter thus:

> How will the struggle for existence act ... in regard to variation? Can the principle of selection, which we have seen is so potent in the hands of man, apply in nature? .... Let it be borne in mind in what an endless number of strange peculiarities our domestic productions ... vary; and how strong the hereditary tendency is. .... Can it then be thought improbable, seeing that variations useful to man have undoubtedly occurred, that other variations useful in some way to each being in the great and complex battle of life, should sometimes occur in the course of thousands of generations? If such do occur, can we doubt (remembering that many more individuals are born than can possibly survive) that individuals having any advantage, however slight, over others, would have the best chance of surviving and of procreating their kind? On the other hand, we may feel sure that any variation in the least degree injurious would be rigidly destroyed. This preservation of favourable variations and the rejection of injurious variations, I call Natural Selection.[202]

Darwin saw that an inevitable consequence of natural selection was the extinction of less improved forms in the struggle for existence. In the fifth edition of *The Origin*, in order to emphasize the idea of improvement, the chapter on natural selection was retitled 'Natural Selection; or the Survival of the Fittest'. The addition of Herbert Spencer's celebrated phrase (often criticized as tautologous) was unfortunate because the adjective *fit* can easily be taken to mean *superior* rather than Darwin's intended meaning of *apt*. The phrase readily leant weight to Spencer's notions of 'Social Darwinism' that emphasised competition, profit, and exclusion of the weaker members of society from its good.[203]

Apart from the question of how variability arises in a species, Darwin was conscious of a number of other gaps and uncertainties in his theory. He knew he would be challenged as to why it was that the fossil records showed no trace of intermediate forms, connecting one species to another. Here he argued that the records were far from complete. He also wondered how natural selection could account for the intricacy of the eye, or how an aquatic animal could gradually adapt to live on dry land. In similar vein he was soon challenged to explain how any incipient organ or characteristic could steadily reach its full development (over many generations) if it conferred no benefit *until* fully developed. How could natural selection operate on the in-between stages? Again, without the later knowledge of the role of genes in the hereditary process, he faced the major question whether a newly emergent trait possessed by a few members of a species would not be rapidly swamped out of existence by the blending effect of inheritance when these members were crossed with the others of the species.[204]

Another major difficulty soon arose, concerning the time scale of evolution. Darwin's theory rested on the assumption of the long time scale suggested by geological research. For example, Darwin estimated that rock strata in southern England were about 300 million years old and he considered this but a moment in geological history. It was therefore with consternation that he learnt of the greatly reduced time scale announced by the formidable Scottish physicist, Sir William Thomson (who became Lord Kelvin). In the 1860s, knowing the rate of increase of temperature with depth below the earth's surface, Thomson calculated the cooling rate of the planet and showed that the latter must have been in a molten state some 100 million years ago – too short a period, it seemed, to have allowed the production of the great variety of species by the slow process of natural selection. It was a sophisticated and convincing calculation, and this estimate was accepted by geologists for nearly fifty years. However, Thomson lacked our 20th century knowledge that the earth has internal processes of heat production[205] which have allowed it to cool far more slowly. It is now accepted that the age of the earth is about 4500 million years, a figure well established from measurements of the radioactive decay of the oldest rocks and from other data.

Darwin took Thomson's assertion seriously. 'I am greatly troubled at the short duration of the world according to Sir W. Thomson', he wrote in 1868. In the final edition of *The Origin of Species* (1872) he faced this problem by assuming with Thomson that the early part of earth history brought more rapid and violent geological changes than those now in progress. Darwin suggested that these would have induced correspondingly faster changes in the organisms which then existed.[206]

Darwin's theory was like a jigsaw puzzle with several pieces missing but with enough of the picture in place to account for a wide range of phenomena. It enabled him to explain, for example, many of his uniquely wide-ranging observations of the geographical distribution of animals. Above all, it provided a reasonable explanation of the countless remarkable examples of adaptation in nature. At times he himself felt uncertain that natural selection constituted the sole evolutionary mechanism – and several of his supporters, including Thomas Huxley, shared that doubt – but he considered it to be the main mechanism. What he had achieved was a remarkable synthesis, even though its key ideas had already received attention separately. It took his genius and patient endeavour to assemble them into a new order and so reveal a structure where others saw only a mass of uncorrelated data.[208]

In the second edition of *The Descent of Man* (1874) Darwin admitted that he had perhaps attributed too much to the action of natural selection. Yet he remained convinced of its central role and its immense potential for the understanding of nature and, indeed, of man himself at both physical and psychological levels. At the end of *The Origin of Species* he wrote:

> In the distant future I see open fields for far more important researches. Psychology will be based on a new foundation, that of the necessary acquirement of each mental power and capacity by gradation. Light will be thrown on the origin of man and his history.

In his own mind the light would illuminate a world of beauty that could no longer be seen as anthropocentric. The book ends with oft-quoted words of a grand vision of the unfolding world of nature:

> There is grandeur in this view of life, with its several powers, having been originally breathed by the Creator into a few forms or into one; and that, whilst this planet has gone cycling on according to the fixed law of gravity, from so simple a beginning endless forms most beautiful and most wonderful have been, and are being, evolved'.

EARLY CONTROVERSIES OVER DARWIN'S THEORY

There is a common belief that remains strangely persistent, even today, that there is a direct conflict between Christian belief and the scientific understanding of the world. This notion grew largely from the kind of championing of Darwin's theory by Thomas Huxley and others as a means to counteract the influence of the Church of England, seen as unduly conservative and dogmatic. Huxley himself sought to promote the idea of the 'church scientific' with the scientists as its high priests, worshipping at the altar of naturalism. The conflict was promoted by two books of the late 19th century, J W Draper's *History of the Conflict between Religion and Science* (1875) and A D White's *A History of the Warfare of Science with Theology in Christendom* (1896). Since then the idea of direct contradiction between the two disciplines has often been fostered by the media in their predilection for combative debate.[209]

Recent scholarship, on the other hand, has revealed a far more complex and interesting interaction. There was no concerted organizing into separate science and religion camps. On the contrary, a large number of learned men – some scientists, some theologians, some who were both, all of whom were religious – experienced various differences among themselves. Altogether the debate was more subdued and unpolarized than generally supposed.

What were the questions raised by *The Origin* in its early years, and what were the main responses? We consider first the concerns on the scientific side, and then turn to the sociopolitical and theological aspects – the latter concern the implications for the stability of society on the one hand, and the credibility of Scripture and the notion of beneficent divine action on the other.

At the outset there was considerable questioning of the substance of Darwin's theory – a theory which assumed that evolutionary change occurred gradually, over vast stretches of time, with natural selection as the chief mechanism. Apart from the doubt raised by Lord Kelvin's estimate of the age of the earth – a mere 100 million years rather than the much longer period assumed by geologists – it was not clear that variations within a species could accumulate to the extent claimed by Darwin, for even within the highly selective process of artificial breeding the degree of change achieved was limited. Besides, the long established idea of the fixity of species died hard. Nor, was it entirely plausible that natural selection was the sole or chief mechanism responsible for evolutionary change. Thomas Huxley, for example, was passionate in his support for Darwin, especially for the latter's naturalism, but he also supported the claim of Darwin's stepcousin, Francis Galton, that evolution occurs not only by slow gradual change but also by 'a series of changes in jerks', in response to extremes of living conditions. As already mentioned, other critics pointed to the liklihood of the swamping of useful variations through interbreeding with members of the species which lacked them, and to the absence of supporting evidence in the fossil record.

But it was Darwin's way of constructing his theory that seemed to arouse the strongest criticism from scientists. The accepted way of doing science was that urged by Francis Bacon two centuries earlier, that is, to gather facts widely and then let them suggest a generalization – a law of nature inferred directly from the data. This *inductive* method was regarded in Darwin's time as the way to obtain certain knowledge of the material world. Fellow scientists felt that Darwin had ignored this path and turned to rather wild hypothesising. He was offering an ambitiously far reaching model without producing any hard evidence of the transmutation of species. The critics wanted proof.[210]

What then was Darwin's approach? He was certainly a patient observer of nature,[211] but not simply a fact gatherer. His path toward natural selection consisted of a complex and highly creative process in which ideas were synthesized into a general hypothesis which was then tested in various ways against his array of observations and experimental data. Thus he followed to a large extent the so-called *hypothetico-deductive* method, which is used in much of modern science. This involves crucially the scientist's well-informed imagination to create the hypothesis and then deduce tests of its validity.[212]

On the socio-political front the idea of descent from the apes, clearly implicit in *The Origin*, caused immediate concern as to how it would be received by Victorian society, restless in its overcrowded cities. The wife of the Bishop of Worcester is reputed to have exclaimed 'Descended from the apes! My dear, let us hope it is not so; but if it is, that it does not become generally known', and the Daily Telegraph spoke of the possibility of 'disastrous consequences to the national peace'. In Germany Darwinism was eagerly received by those with ideas of liberal reform and even social revolution. But if it was taught in school classrooms, others wondered, would it not raise a generation 'whose confessions are atheism and nihilism and whose political philosophy is communism?'[213]

Then there was the anxious question whether brutish descent implied brutish behaviour. Perhaps it was mainly this that prompted Charles Lyell, otherwise a friend and supporter of Darwin, to resist seeing the human race as simply a natural product of the evolutionary process, for he emphasised consistently the uniqueness and 'high genealogy' of our species. Again, in a letter to Darwin, Adam Sedgwick wrote that he greatly admired certain parts of *The Origin* but had read other parts with deep sorrow because he thought them 'utterly false and grievously mischievous' – Darwin had given no place to the link between the realms of the moral and the material.[214]

The broad range of responses to *The Origin* are surveyed comprehensively by James Moore in his book *The Post-Darwinian Controversies* (1979). Those who responded in terms of a Christian world-view he groups under the labels: anti-Darwinian, Darwinian, and Darwinist. The Darwinian and Darwinist groups – which, roughly speaking, may be labelled theologically *orthodox* and *liberal* respectively – both welcomed the publication of *The Origin*, but for different reasons which we consider below.[215]

For the Christian anti-Darwinians, an obvious point of conflict lay in questions of biblical authority and exegesis; the new theory was clearly incompatible with a literal interpretation of the creation accounts in Genesis and with the general world-picture shaped by them. It should be noted, however, that whatever Darwin's impact on Christian confidence in the Bible, the much greater disturbance came from the rise of historical and literary criticism – the so-called 'higher criticism' of Scripture. Alan Richardson refers to *The Origin of Species* as only a side-issue in what has been termed the Victorian crisis of faith. The real issue was not between Genesis and Geology or between Darwin and the Bible, but between the traditional and the critical approaches to biblical-historical study and interpretation.[216]

At the British Association conference in 1860, the Bishop of Oxford, Samuel Wilberforce, not only presented objections on scientific grounds but, in his celebrated clash with Thomas Huxley declared that the principle of natural selection is absolutely incompatible with the word of God – and from the Episcopal Church in the USA came the following edict, 'if this hypothesis (evolution) be true, then is the Bible an unbearable fiction'.[217] The conservative position on the authority of Scripture was put trenchantly by the Oxford theologian J W Burgon (who rejected Darwin's theory vehemently as contrary to Scripture):

> The Bible is none other than the voice of Him that sitteth on the throne. Every book of it, every chapter of it, every verse of it, every word of it, every syllable of it (where are we to stop?), every letter of it, is the direct utterance of the Most High. The Bible is none other than the Word of God, not some part of it more, not some part of it less, but all alike the utterance of Him who sitteth upon the throne, faultless, unerring, supreme.

For some Christian anti-Darwinians the main theological objection to Darwin's theory lay in the perception that it undermined the idea of design in nature, thus bringing into question the very existence of God. For some the turmoil was enough to bring 'agonies of terror' and even the loss of religious belief altogether. There were many Victorians, steeped in Christian tradition, whose manuscripts and memoirs reveal a common struggle with the ideas and implications of Darwinism.[218]

Darwin himself felt that there seemed to be no more design in the variability of organic beings, and in the action of natural selection, than in the course which the wind blows.[219] Yet, in keeping with Victorian optimism that the course of history is a continuing story of human progress, he could also write in the penultimate paragraph of *The Origin of Species*: 'As natural selection works solely by and for the good of each being, all corporeal and mental endowments will tend to progress towards perfection'.

In the USA the anti-Darwinian theologian Charles Hodge (1797-1878) – authoritative teacher at Princeton Theological Seminary, highly effective controversialist, and a discerning critic from the standpoint of conservative Christianity[220] – focussed his attack on the suggestion of randomness in the natural selection process, an idea that seemed to allow no role for Providence. The problem from the point of view of Christian belief was that natural selection came to be seen as a rigorous naturalism in which no external agency was required either to direct the course of development or to guarantee that a transcendent purpose was being realized as species changed over time.[221] So, whatever Darwin's own particular state of agnosticism, his theory seemed *effectively* atheistic. In his final book *What is Darwinism?* (1874), Hodge provided the uncompromising answer: It is atheism. His stand was backed by the lesser voices of many clergymen and not a few academic leaders.

On the other hand there was notable support for Darwin from both theological wings – from the 'orthodox' Darwinians and the 'liberal' Darwinists. Both groups sought a reconciliation between the new theory and Christian belief. The former group found in Darwinism no contradiction of their orthodox doctrines and were able to support it wholeheartedly, seeing it as a convincing explanation of the created order – a creation burdened with struggles and imperfections, yet sustained by an immanent all-caring God. On the other hand, while supporting the idea of evolution and even natural selection, the Darwinists were keen to avoid Darwin's notion of 'chance' variations, for this seemed to imply a lack of direction in the evolutionary process. Instead they were inclined to invoke either the Lamarckian influence of the environment or the divine impress of teleological 'directivity' in organic material – only thus would the idea of evolution match their belief in a world of progressive development.

Among the early Darwinist voices was that of Frederick Temple (1821-1902) – contributor to *Essays and Reviews* and enthroned as Archbishop of Canterbury in 1896 – who spoke for many in his interposing of divine agency as the primary cause of evolution. He did not believe that Darwinism affected the substance of Paley's argument from design, for natural selection was only one element in the overall process. In his 1884 Bampton Lectures he maintained that the Creator 'impressed on certain particles of matter . . . such inherent powers that in the ordinary course of time living creatures such as the present were evolved'.[222]

James McCosh (1811-1894) – president of the College of New Jersey which later became Princeton University – expressed a liberal attitude towards science, declaring: 'We give to science the things that belong to science, and to God the things that are God's. When a scientific theory is brought before us, our first enquiry is not whether it is consistent with religion, but whether it is true'. He gave a large place to Darwinism as a biological explanation of the emergence of *homo sapiens* but thought that special divine action may have been needed for the 'breath of life' referred to in Genesis – and needed, certainly, in the fashioning of the human soul.[223]

One of the remarkable members of the Darwinist group was the Scottish naturalist and Free Churchman, Henry Drummond (1851-1897). As an ardent disciple of Herbert Spencer he looked to combine Christian belief with the philosophical idea of evolutionary progress. But he took the idea to an extreme level, applying it not to human social development but rather to the spiritual progress of humankind. Indeed, he reached a point in his final book *The Ascent of Man* (1894) where he seemed so to stress the role of altruism in the late stage of biological evolution that he could speak of Christianity and Evolution as one and the same thing. He wrote: 'What is Evolution? A method of creation. What is its object? To make more perfect living beings. What is Christianity? A method of creation. What is its object? To make more perfect living beings'. Thus, Evolution/ Christianity is not only progress in matter, it is progress in spirit, 'the phenomenal expression of the Divine, the progressive realization of the Ideal, the Ascent of Love', an all-embracing creed of progress. Near the end of his short life he faced a barrage of critical comment and even a charge of heresy, of which he was acquitted.[224]

Of the Christian Darwinians, the American botanist Asa Gray (1810-1888) was one whose support was especially valued by Darwin himself. He was a professor of natural history at Harvard University, author of the pamphlet *Natural Selection not Inconsistent with Natural Theology*, and very well known in the USA. His concern was to show the process of evolution as itself the object of design by the Creator. With the causes of variation in species being unknown, Gray made the assumption that the hidden hand of God was in the evolutionary process – a notion that lay open to criticism as a 'God of the gaps' explanation. Darwin felt obliged to respond: However much we may wish it, we can hardly follow

Professor Asa Gray in his belief that variation has been led along certain beneficial lines, like a stream along definite and useful lines of irrigation. Gray's brave attempt to bridge the divide between the domain of science and the domain of both metaphysics and theology, as Darwin's niece Julia Wedgwood pointed out, was not strong enough to overcome the polarized emotions of those who rejected religious reality *in toto* and those who fell back on tradition with a conscious rejection of science.[225]

Charles Kingsley (1819-1875), poet and the author of *The Water Babies* and other tales, was another close supporter of Darwin. He produced several scientific and theological works and was drawn to the Christian Socialism of the London theologian F D Maurice. He described himself as an old-fashioned High Churchman, an orthodox priest of the Church of England who believed its theology to be eminently rational as well as scriptural. As an enthusiastic naturalist he was deeply impressed by the pre-publication copy of *The Origin* that he received from Darwin, replying that 'if you be right, I must give up much that I have believed and written'. Nevertheless he was clear that there was no contradiction between evolutionary theory and Scripture: 'Scripture says that God created. But it nowhere defines that term. The means, the How of Creation is nowhere specified'. Nor did he consider the argument from design to be undermined by the theory. One could accept all of Darwinism and yet preserve 'our natural theology' on exactly the same basis as that on which Paley left it, even though it might need to be developed. He summarized vividly his new theological view:

> We knew of old that God was so wise that He could make all things; but behold, He is so much wiser than even that, that He can make all things make themselves.[226]

The Scottish theologian James Iverach (1839-1922) was another who claimed that Darwin's theory may be held in such a form as to have no dangerous consequences for philosophy or theology. He took honours in the mathematical and physical sciences and in due course, after theological training and thirteen years of parochial ministry, was appointed to the chair of apologetics in the Free Church College, Aberdeen. Here he made his mark in defending and restating Christian truth, faced by the claims of naturalism and other forms of speculative philosophy. One appreciative student wrote that 'the brightness of his

Christian faith, inherited and then made his own, could not be disintegrated by the ever-growing complexities of the Spencerian evolution, nor obscured by the dazzling cloudbanks of Hegelian speculation'.[227]

Iverach could readily come to terms with an 'evolution' which begins and proceeds in accordance with free and rational purpose. What was not admissible was the idea that evolution could get 'the determinate from the indeterminate, intelligence from the unintelligible, something from nothing' – without intelligence there could be no order. He explained that such a claim does not stand as an alternative to a scientific conception of the world but as an interpretation of scientific findings from a religious point of view. When viewed in this wider context, science is strengthened by the religious conviction that nothing occurs by chance.[228] Conversely, he continued, the argument from order to intelligence is much more cogent than it was in Paley's time for 'no one ever strengthened the argument as Darwin has done'. But without an underlying purpose in the evolutionary process, natural selection can produce nothing by itself, it can only eliminate the unfit. The real question then is the origin and nature of the variations selected. What was needed was to search for the unknown laws of variation and when discovered, he claimed, they will give evidence of God's working out His purposes, delivering us from the tyranny of chance. Sounding a trinitarian note of the 'creative and sustaining activity of the Logos' he suggested that

> Modern science and philosophy will have done us a great service if it will force students of theology to go back not merely to the history of theology, but to the New Testament itself, to search out its meaning, to gather together and to set forth in order and method its profound teaching on the relation of God to the world.

In the late 19th century it fell to the Anglo-Catholic theologian and amateur botanist at Oxford University, Aubrey Moore (1843-1890), to play a large role in the breaking down of antagonisms toward evolution still widely felt in the Church of England.[229] Being also the curator of the University's botanical gardens he was equally at ease among scientists and theologians and was well placed to serve as a link between the two groups.

At a time when Christian apologetic tended to see the creative work of God in the gaps left by the processes of nature – especially the gap between inorganic matter and life, and the gap between rational man and all the other animals – Moore, a forceful Darwinian, criticized this view fiercely. Christian belief should not be pinned to the assumption that the gaps would never be closed. So he refused to connect the Christian faith either with evolution or with the denial of evolution. Whatever science revealed in this area was, after all, only a revelation of God's method of creation – 'evolution *or* creation' he regarded as a false antithesis. If evolution is indeed God's way of creating, Christians should think of it as 'supernatural evolution' or 'natural creation' – otherwise one acquiesces in a deism that separates the natural and the supernatural. Then divine activity in the world would be simply a series of supernatural interferences in the course of nature. Indeed for the Christian theologian, he claimed, the facts of nature *are* the acts of God.[230]

With this intense emphasis on divine immanence, Moore found it difficult to accept the claim of both Charles Kingsley and Frederick Temple that God in his great wisdom makes things make themselves. This seemed to imply the deistic picture of a God who withdraws himself from his creation, and leaves it to evolve itself, though according to a foreseen and fore-ordered plan. Moore felt that to speak of a world imbued with a power of self-unfolding was to contradict belief in God.

He went on to criticize the widely held belief in the *special creation* of humankind. The latter was without biblical, patristic, and medieval authority, he asserted – it had arisen as part of 17th century natural theology, having received a great impetus at that time from the Puritan poet, John Milton, in his *Paradise Lost*. Moore regarded the Darwinian picture as far more Christian than the theory of special creation because it implied the immanence of God in nature and the omnipresence of his creative power. Those who defend special creation and other acts of intervention by God seem to have failed to realise, he wrote, that *a theory of occasional intervention implies as its correlative a theory of ordinary absence*, fitting the deistic picture of God as absentee landlord.[231]

As Moore considered the Christian doctrine of God in the light of Darwin's theory, he felt that it had been enriched and renewed. It had helped the Church to recover an understanding of God's triune nature that had been distorted by the deism and semi-deism of the Enlightenment. He wrote:

> Science had pushed the deist's God further and further away, and at the moment when it seemed that He would be thrust out altogether, Darwinism appeared, and, under the guise of a foe, did the work of a friend. It has conferred upon philosophy and religion an inestimable benefit, by showing us that we must choose between two alternatives. Either God is everywhere present in nature, or He is nowhere .... In nature everything must be his work or nothing.

It seems as if in the providence of God, he added, the mission of modern science was to bring home to our unmetaphysical ways of thinking the great truth of the Divine immanence in creation.[232]

But how is that immanence to be construed? In other words, what can be said about the nature of *creatio continua*, the unceasing work of divine action in the world? – a question that has continued to occupy the minds of theologians and others in today's Science and Religion enterprise.

A century later the consequences of Darwinian theory and other major developments in science (such as the decoding of the human genome) can appear very different. The following quotation is perhaps typical of the common trend to regard the progress of science as the steady outdating of religious belief, certainly in the western world:

> In modern science's culminating triumph over traditional religion, Darwin's theory of evolution brought the origin of nature's species and man himself within the compass of natural science and the modern outlook ... a particular individual's outlook in the modern era could occupy any position in a wide spectrum from a minimally affected childlike religious faith to an uncompromisingly tough-minded secular skepticism.[233]

However, with the unfolding of twentieth century science, together with its application in both constructive and destructive ways, some basic questions about the universe and human existence have arisen afresh. One notable response has been the emergence of the phenomenon of Christian fundamentalism – as a rallying call to the 'fundamentals' of the faith, especially in the USA of the 1920s, and then as a world-wide campaign of the Scientific Creationism movement since the 1960s to re-introduce the idea of 'special creation' into the public arena, especially into school science syllabuses.[234] Within some circles of liberal and evangelical theology on the other hand, there has developed since the 1960s a new interest in natural theology, in which fresh consideration is being given to the argument for the existence of God from the pattern of the universe as a whole, and even to the question of constructing a new metaphysics from the insights of both theology and the sciences.

THE ELABORATION OF DARWIN'S THEORY

By the end of the 19th century, Darwin's primary idea was close to collapse. While there was almost universal acceptance of the longstanding idea of *evolution* among scientists, there was widespread doubt among biologists about natural selection as the main mechanism, especially in the absence of a sound theory of heredity. Some were drawn to neo-Lamarckian assumptions of the inheritance of characteristics – either acquired through frequent use or directly induced by the environment – but there was no clear consensus.

Then came the breakthrough of Mendelian genetics when the obscurely reported work of Gregor Mendel in the 1860s was discovered and taken seriously by the Dutch biologist Hugo de Vries in the year 1900. Julian Huxley (grandson of Thomas Huxley) has written of the eclipse of Darwinism that took place at that time, before its revival in combination with the new genetics.

The subject of modern genetics arose early in the 20th century when it was suggested that Mendel's hereditary factors or 'genes' are located in the nucleus of the reproductive cells. There they are linked together to form thin threads called chromosomes, which are observable under a microscope. The study of genetics has provided the answer to the question about the swamping of favourable characteristics, showing that the assumption in Darwin's time that offspring inherit the *mean* of their parents' traits is wrong. In spite of backcrossing with the majority of the species, a new property can indeed be passed on to successive generations without being steadily diluted. Again, since Darwin's time, it has become clear that the existence of variations within each species, on which the process of natural selection depends, arises from mutations of the underlying genes and chromosomes – by means of x-rays and other types of radiation, and by chemicals, heat and other factors. Variations arise too from the phenomenon of genetic drift, that is, the slightly inexact copying of genes in the transmission from parents to offspring.

In the early decades of this century it was suggested that the mutations of genes can lead to saltations or 'jumps' in the development of a species. The gradual accumulation of small differences – the process advocated by Darwin – was relegated to a peripheral role. Paleontologists on the other hand were increasingly confident from the fossil record that Darwin was right. Then within the period 1936-47 the most viable components of the previously competing research traditions were brought together and scientists who had previously held different views about evolution came to a common mind, achieving what became known as the 'neo-Darwinian synthesis'. This modified theory has been summarized recently by R J Berry.[235] In spite of continuing arguments about some aspects, this synthesis of Darwinian theory and modern genetics constitutes the framework within which biologists operate. For example a leading neo-Darwinist, Theodosius Dobzhansky, has made use of this fundamental assumption as the title of a paper: 'Nothing in biology makes sense except in the light of evolution'.[236]

In his book *God and Evolution* (1988), Berry has described thus the significance of neo-Darwinism for biology:

> The importance of the neo-Darwinian synthesis is that it re-established the unity of biology which Darwin's ideas had originally provided, and thus made possible generalisations within an otherwise impossible diversity of living organisms. The Periodic Table gives a similar service to chemistry. It is this unifying element which apparently makes evolution into something more than a simple scientific theory, and allows such diverse topics as fossil sequences, gene frequency changes and polymorphism, extinctions, adaptation, and so on, to be brought within a single umbrella. There may be disagreement about the interaction or relative

> importance of particular mechanisms, but there is no viable alternative to Darwinian evolution for understanding nature.[237]

At the centenary of *Origin of Species* in 1959 very few doubts were expressed about the overall neo-Darwinian synthesis, in which natural selection continued to be regarded as playing the dominant role in the processes of evolution. As Berry mentions, however, disagreements have since arisen within biology concerning the relative significance of various evolutionary mechanisms,[238] prompted by increased knowledge and understanding. Adherents of Scientific Creationism look to such questioning as support for the overthrow of 'evolution', but for biologists the basic idea of the development and transformation of species remains axiomatic. Mainstream biology still gives a central place to natural selection.

Evolutionary development is by definition a gradual process, but one would expect the pace to change during or after the occurrence of extremes of living conditions, as in an ice age. This is the point stressed in the paleontologically inspired picture of 'punctuated equilibria'. Here earlier life-forms have persisted virtually unchanged throughout long periods (millions of years) and then have changed quite suddenly (over a few tens of thousands of years).[239]

For example, marine bivalves (small clam-like creatures) existed in the Jurassic Period (between 180 and 135 million years ago) remaining almost unchanged for periods of 10 million years or more and were then replaced quickly with a markedly different species – sufficiently quickly to account for the absence of intermediate forms in the fossil record. The main point in the idea of punctuated equilibria is that the sudden evolutionary changes coincide with major environmental catastrophes which cause mass extinctions. The species which survive then find themselves in an environment with less competition for the resources and with many empty environmental niches into which they can expand – ideal conditions for 'adaptive radiation'[240] into those niches. Unlike the catastrophism of the early 19th century, this 'neo-catastrophism' makes no particular claim to divine action in the adaptive radiation process.

Research in evolutionary biology is now concerned not only with the development of a given species but also with the co-evolution of a variety of species in an ecological system. But perhaps the most important new concept in this area is the notion of self-organization in nature, discussed later in the section on *complexity theory*. This is the propensity of complex systems far removed from equilibrium to form themselves into structures of increased order.

What was Darwin's legacy? As one who studied widely, he was remarkably successful in making accessible and understandable to a broad spectrum of readers evidence that ranged over several disciplines – geology, botany, taxonomy, and morphology for example. The fertility, creativity and accessibility of his work has meant that its influence soon spread beyond the concerns of natural science to make its mark in philosophy, the social sciences, and Victorian literature, and it raised fundamental questions for theology and Christian belief. Altogether it continues to inspire a broad range of academic endeavour – a veritable Darwinian industry.

4 THE EVOLVING UNIVERSE: NEW INSIGHTS

CLASSICAL PHYSICS

The culmination of early modern science in the publication of Newton's *Principia* (1687) was followed by a century of modest development in physics. Then came a number of significant advances in the 'classical physics' of heat, light, and electricity: the demonstration of the wave nature of light (1801); the discovery of the magnetic effects of electric currents and the electric effects of moving magnets (1821-31), which in due course formed the basis of electric motors and dynamos; a new understanding of the nature of energy and its transformation from one type to another, culminating in the celebrated Second Law of Thermodynamics[241] (1850); and Maxwell's elegant theory of the propagation of electromagnetic waves (1864), which established the profound link between electricity and magnetism and showed that light and radio waves belong to the same electromagnetic spectrum.[242] Altogether it was a flowering that undergirded the industrialization and urbanization that grew in the Western world from the early 19th century onwards.

Towards the end of the 19th century so impressed were physicists by these developments that some began to think that all the important laws of physics had been discovered and that research would henceforth be concerned with merely clearing up minor problems. Few could foresee that the world of physics was on the eve of epoch-making discoveries, destined on the one hand to stimulate physics research as never before and on the other to usher in an era of the application of physics to industry on a scale previously unknown.

THE NEW PHYSICS

Since 1900 science has created a new world-picture and led to a vast array of applications, both creative and destructive.

> The fruits of science have irrevocably changed the way we live ... for modern mathematically based physics has spawned technologies that have altered the very fabric of daily life. Electric power, radio and television, the internal combustion engine, the airplane, the telephone, the silicon chip, lasers, and optic fibers are all by-products of this science. During the past century, physics-based technologies have redefined how we work, play, entertain ourselves, and communicate with one another. At the same time, these technologies have given us unprecedented destructive power: laser-sighted guns, guided missiles, supersonic fighter planes, nuclear-powered submarines, and, of course, atomic bombs. Physicists' discoveries have changed our lives as powerfully as any political, economic or religious forces. Along with universal suffrage and parliamentary democracy, physics is one of the primary forces that have shaped Western culture[243].

Underlying much of this development was the dramatic advance in theoretical physics during the first three decades of the 20th century, especially in the understanding of the world of atoms and sub-atomic particles through the techniques of quantum theory. The insights of atomic physics led in turn to the evolution of chemistry from a vast but rather formless body of empirical knowledge into a coordinated science. Again, ideas and techniques in physics soon contributed to the new science of genetics, with its key discovery of the structure of the DNA molecule in 1953. This in turn formed the basis of genetic engineering and the vast undertaking to map the genetic makeup of the human body, the Human Genome Project. Quantum theory was also a key to the invention of the laser in the mid-20th century, with its wide variety of applications, and to much of the development of the physics of condensed matter, especially the discovery of the transistor.

In the area of technology, multiple arrays of transistors in the form of miniature silicon chips led to increasingly compact and powerful computers. These have since fueled many of the advances in science and technology through their capacity to receive and process large amounts of numerical data. For example, they play an essential role in the operation of particle physics experiments, in the control of chemical manufacturing processes and spacecraft missions, in the guiding of delicate manoeuvres in brain surgery, and in the mapping of the cosmos with large telescopes. Furthermore, because computers are capable of vast numbers of iterations of an arithmetical instruction, they have made feasible the simulation of *nonlinear systems*, that is, systems of interlinked components showing great complexity in their unfolding process. This has contributed greatly to the understanding of systemic behaviour, both physical and biological.

Taken together, the realms of the very small (particle physics), the very large (astrophysics and cosmology) and the very complex (nonlinear dynamics or dynamical systems theory) constitute what has been called The New Physics.[244] We survey each of these briefly.

PARTICLE PHYSICS

The Newtonian ideas of absolute space and absolute time were profoundly modified in Albert Einstein's theory of Special Relativity

(1905). With the axiom that the speed of light is the same for all observers, irrespective of their motion, he showed that time and space intervals are seen differently by any pair of observers in relative motion. From this followed his celebrated statement of the equivalence of mass and energy, expressed mathematically as $E = mc^2$ (where c represents the speed of light). Apart from its philosophical interest, this equation has had immense implications for the understanding and use of nuclear energy.

The ideas of Special Relativity, appropriate for observers with unchanging velocity, were extended by Einstein to the more general case of accelerated motion – and the resulting theory of General Relativity (1916) provided a new way of understanding the operation of gravity in the universe. One of its predictions was that light itself, as a form of energy and therefore possessing the property of mass (however small), is attracted gravitationally by another mass; for example, a beam of starlight should be bent slightly as it passes close to the sun. The theory was widely acclaimed when such bending was sought and observed on the occasion of the solar eclipse of 1918. General Relativity has since provided the foundation for a much more rigorous approach to the making of cosmological models from astronomical observations.

A new view of the microworld of particles began to develop in 1900 when Max Planck suggested that light – a form of wave motion – is emitted in multiples of a small unit or 'quantum' of energy which is proportional to its frequency.[245] Five years later Einstein took this idea further, suggesting that light travels as a stream of discrete particles or 'photons', thereby explaining certain puzzling features of the emission of electrons from light-irradiated surfaces of metals. This led to Niels Bohr's simple planetary model of the atom (with electrons revolving around a positively charged nucleus, like planets around the sun) and in due course to an extremely successful quantum theory of particle behaviour. The theory appeared in two equivalent forms, invented by Werner Heisenberg and Erwin Schrödinger in 1925, which have been developed and refined over several decades to deal with different aspects of the very small.

One of the early breakthroughs was the discovery of the neutron in 1932, protons and neutrons being the building blocks of the atomic nucleus. Thereafter the combination of more and more powerful particle accelerators and increasingly sophisticated quantum theories has led to the discovery of a veritable 'zoo' of new elementary particles. These comprise the various leptons (such as the electron and neutrino), quarks (such as the up and down quarks that are the building blocks of protons and neutrons), and force-mediating messenger particles[246] – photons for the electromagnetic force that holds atoms and molecules together, gluons for the strong nuclear force that binds the protons and neutrons in the atomic nucleus, and the massive W and Z particles for the weak nuclear force that plays a crucial role in radioactive decay, supernova explosions, and the nuclear burning processes of the stars. Gravity is not included in this list because it is significant only for much larger masses.

A key feature of the microworld is the stability of not only the common members – especially electrons, protons and neutrons[247] – but also their arrangements in the different types of atoms. These constitute sharply defined chemical elements in each of which the atoms are all identical. The significance of such stability is that it allows the evolution of stars and biological forms to take place over millions of years. Furthermore, in the continuous chemical processes in living things, many of the atoms of any given chemical element are being gradually replaced by others of the same element, and the fact that the old and new are identical (as well as stable) helps to guarantee the stability and continued identity of each living thing over its lifetime.

Two fundamental aspects of the microworld have emerged. The first is known as Heisenberg's Uncertainty (or Indeterminacy) Principle, which specifies a limit on how precisely one can know *both* the position and velocity of a particle (at any moment). The more precisely either of these variables is determined, the less precisely can the other be known. Although initially Heisenberg took this to be a diagnostic limitation, it is usually regarded as a statement about inherent indeterminacy, that is, an assertion of ontological rather than epistemic indeterminacy. The second aspect of the quantum world is a property of particle interactions known as 'non-locality', perhaps better designated as 'quantum entanglement'.[248] This refers to a certain strange ongoing connectedness between particles that are linked and then separated. Once they have been linked a change in one takes effect in the other within a time interval less than the time for a signal to travel between them at the speed of light, that is, there seems to be *immediate* communication. Quantum theory introduces us, too, to an elusiveness in the nature of the decay of a particle into two or more new particles – an event that just happens, seemingly uncaused.

These features of the micro-world are counter-intuitive and make it difficult to picture particle events, even though they may be well described in terms of mathematics.[249] At the level of its constituent particles the physical world is characterized by a certain fuzziness and fitfulness that contributes to the view that the world is essentially nonmechanistic and open to novelty in its development. Altogether the increasing grasp of the strange world of particle behaviour constitutes the greatest revolution in physics since the time of Newton. It is a revolution that has continued to mature over several decades and, in the form of the 'supersymmetric string theory' of the past few years, is promising to develop into a fully unified description of the fundamental particles and forces, including gravity.[250]

COSMOLOGY

With advances in the understanding of particles have come discoveries in astronomy which have sparked a new view of the universe itself. In the early 1900s, in spite of the 'changes and chances of this fleeting world', the celestial realm seemed settled and static. Sun, moon and stars seemed to shine unendingly in their ordained positions. The universe had been mapped with increasing accuracy throughout the previous two centuries (especially with the advent of photographic techniques in astronomy in 1850) and many faint patches of light had been seen in all directions except those in the plane of the Milky Way.[251] Immanuel Kant (1724-1804) and others speculated that these patches were actually separate galaxies, or even 'universes', but most astronomers disagreed; the fact that they are indeed separate galaxies in our immense universe did not become well established until the 1920s. The realization, only a few decades ago, that our galaxy is not unique and central in the universe ranks as one of the major advances in cosmological thought. But there was far more to come.

In the late 1920s Edwin Hubble discovered that the galaxies are moving away from each other. On closer examination he found that the recessional speed of a galaxy is proportional to its distance away, as though the entire space of the universe were uniformly expanding from some initial compact state, carrying matter with it.[252] Furthermore, from the observed speeds it was a simple matter to calculate roughly how long ago the universe was in its highly compact state – a state that marked, perhaps, the beginning of it all. The calculated age of our expanding universe has varied between ten and twenty billion years and is often said to be about thirteen billion years, that is, about three times the age of the earth.[253] The powerful Hubble Telescope now orbiting the earth is detecting light from quasi-stellar objects, or quasars, in the far reaches of the visible universe. In the case of the most distant quasars and galaxies yet observed, the light has been travelling through space for more than ninety per cent of the age of the universe, revealing something of those bodies *as they were then.*

Hubble's discovery gave rise in the late 1940s to two rival theories of the expanding universe. Hermann Bondi, Thomas Gold and Fred Hoyle suggested that the universe is always and everywhere the same, with new matter continuously created throughout by some sort of 'creation field' which keeps the density constant.[254] This simple and, in some ways, aesthetically pleasing Steady State model avoided the problematic question of the singular event of a beginning: how to explain and describe such a moment where the laws of physics break down. As a bonus for Hoyle the model seemed also to dispense with the need for a creator.[255]

The rival picture was one in which the universe started with what Hoyle derisively termed a Big Bang, expanding explosively from an unimaginably hot compact state which marked the beginning of time and space, and of all matter and its associated energy.[256] Cosmologists have considered the possibility that the Big Bang was preceded by a phase of cosmic contraction that culminated in a Big Crunch, from which the universe then bounced back. However, there are no known laws of physics that can be applied to the ultra-extreme conditions of the very earliest moments, so the Big Bang itself is inherently inaccessible to the imagination of theory-building physicists. Nevertheless, speculation continues and the picture of an endlessly expanding and contracting universe has appealed to some. Similarly, the idea of our visible universe as simply one expanding bubble amongst countless other such bubbles in a much vaster region has also been considered. But these ideas of the *before* or the *beyond*, although conceivable, are inherently untestable and therefore must be regarded as speculations lying outside the realm of science itself – as candidates for some metaphysical scheme perhaps.

In 1948 George Gamow suggested that the remnants of the initial flash might fill the universe even now, like the radiation filling an oven, but cooled by the expansion to within a few degrees of absolute zero (-273 degrees C). In 1965 the extremely faint cosmic background radiation was stumbled upon during the testing of an ultra-sensitive radio telescope in the USA. A more precise set of data obtained in 1991 from the Cosmic Background Explorer satellite[257] showed (to an accuracy of better than 1%) that the present temperature of the universe is 2,73 degrees above absolute zero. A further triumph was the discovery that the relative amounts of the primordial atoms[258] – the hydrogen, helium, and traces of deuterium and lithium which have persisted from the time of the early universe – are approximately the values calculated from the temperature.

This is a provisional theory which hangs together well, drawing upon knowledge of particle physics to describe the earliest moments of the cosmos. It is the present standard model in cosmology, no longer rivalled by the Steady State theory. It allows physicists to work backwards in time and determine, stage by stage, the thirteen billion year history of the universe, at least to within a few minutes of the Big Bang. Beyond that the account is boldly speculative but remains coherent to within a trillionth of a trillionth of a trillionth of a second of the Big Bang, at which stage the absolute temperature must have been higher than it is now by a factor of some thirty orders of magnitude.[259] The universe would have cooled through most of that range in the first hundredth of a second, allowing the quarks to begin to cluster into protons, neutrons and other composite particles. After a few minutes the lightest nuclei would have been able to hold together, and after a few hundred thousand years the same would apply to atoms.[260] But it is in only the late stages of cosmic history that conditions have been appropriate for the forming of the complicated molecules needed for life, however this process may have occurred. A rough outline of the chronology of the universe is given on the next page.

Cosmic evolution is now pictured as follows: After the early formation of the primordial atoms (mostly hydrogen and helium) there followed a long comparatively uneventful period in which gravity came inexorably to dominate the large-scale scene, condensing matter into clusters of hot dense stars with their life-cycles of burning and radiating. The process of *nuclear fusion*[261] in stellar interiors converts hydrogen into helium. In the large stars helium then burns in the same way to form successively carbon, oxygen, and eventually, through this and other nuclear reactions, the entire range of chemical elements. The fusion process is accompanied by the emission of heat, light and particles, as in the case of the sun's burning and radiating. After several billion years some of the larger first generation stars exploded as supernovae, making the contents available for second generation stars and their planets.

Table 2. Chronology of the Universe

time after Big Bang	temperature of universe[262]	occurrence
0		Big Bang
10^{-43} secs	10^{32} K	forces all equal matter = antimatter
10^{-33} secs	10^{27} K	matter > antimatter gravity weakens
10^{-9} secs	10^{15} K	weak force weakens
10^{-2} secs	10^{11} K	quarks begin to cluster
10^{2} secs	10^{9} K	lightest nuclei form
10^{5} years	10^{3} K	atoms form, light travels freely
much later		stars and galaxies form, then chemical elements in stars, some of which explode in due course as supernovae;
10^{10} years		new stars and solar systems; earth forms; then after 1 billion years, micro-organisms;
	3 K	after another 2 billion years, increasingly complex forms

What of the future of the universe? Will it always continue to expand or does it contain enough mass to draw itself back into an equally long period of contraction under the action of gravity? At present this is an open question since cosmologists are not sure how much 'dark' invisible matter lies hidden in the galaxies, but the present indications are that there is not quite enough mass in the universe for the force of gravity to reverse the expansion. In that case the destiny, in scientific terms, would be the 'heat death' of unending expansion and cooling.

COMPLEXITY THEORY

As the standard theory of elementary particles emerged and began to find application to the earliest moments of the universe, some scientists were turning their attention to the more complex phenomena in nature – seemingly intractable things like turbulence in the sea or atmosphere, fluctuations in wildlife populations, or the oscillations of the heart or the brain. Powerful computers opened up the study of *nonlinear systems.*[263] These are mechanical, electrical or biological systems of interlinked components, often simple in their structure but capable of great complexity in their unfolding process. Because their dynamical behaviour can become chaotically irregular, the *nonlinear dynamics* of such systems is often called *chaos theory* (or simply *chaos*), even though the latter constitutes only a part of *nonlinear dynamics*, not the whole of it. A more comprehensive name is *complexity theory* which covers not only the aspect of *chaos* but also the phenomenon of *order* that can arise in nonlinear systems far from their equilibrium condition. We consider chaos first.

In his early review of the topic James Gleick writes:

> The irregular side of nature, the discontinuous and erratic side – these have been puzzles to science. But in the 1970s a few scientists in the United States and Europe began to find a way through disorder. They were mathematicians, physicists, biologists, chemists, all seeking connections between different kinds of reality. Physiologists found a surprising order in the chaos that develops in

> the human heart, the prime cause of sudden unexplained death. Ecologists explored the rise and fall of gypsy moth populations. Economists dug out old stock price data and tried a new kind of analysis. The insights that emerged led directly into the natural world – the shapes of clouds, the paths of lightning, the microscopic intertwining of blood vessels, the galactic clustering of stars.[264]

Nature's most complex systems tend to lie somewhere in the middle of the vast range of length scales in the universe, between those of electrons and galaxies. The size of the most complex system we know, the human person, lies at about the geometric mean of the size of an atom and that of a large star. Furthermore, atoms and stars are relatively simple systems, whereas the most complex biological organ we know of, the human brain, has about a hundred million million synaptic switches, and the number of states in which it can exist greatly exceeds the number of atoms in the visible universe. Even a single-cell creature such as the paramecium, which is about a tenth of a millimetre long, is far more complicated in its structure and processes than a galaxy of stars.

Most systems in the natural world are *nonlinear* and, consequently, exhibit complicated behaviour over certain ranges of their operating parameters.[265] As the parameters are changed a system can pass from simple regular behaviour that repeats itself exactly to the highly complex non-repetitive irregular behaviour described as *chaotic*. Between these regimes of stability and chaos there is a transition zone in which the system samples, in orderly fashion, a much wider variety of arrangements or positions than in the stable regime. Here the system is said to be *at the edge of chaos.*

Recent study of evolutionary change via computer modelling of binary networks has shown the importance of the concept of *the edge of chaos*, introduced by Stuart Kauffman in his work at the forefront of the search for the laws of complexity.[266] He suggests that natural selection operates on organisms, whether as an individual species or a co-evolving group of species, so as to bring them to the edge of chaos, where interaction is vigorous but not chaotic – and he sees this as perhaps a general characteristic of evolving biological systems. If the behaviour is too ordered it cannot coordinate the complex sequences of genetic activities necessary for development – and if it is too far into the chaotic regime it lacks the continuity necessary for development.[267]

One of the long standing problems about the neo-Darwinian theory of evolution has been the question of how the present variety of species could have been produced in the time available through *simply* a gradual process of natural selection, even if the process is allowed to stretch over two or three billion years and is optimized at the edge of chaos.[268] So, attempts are being made by Kauffman and others to understand the phenomenon of self-organization in nature as an inherent aspect of evolutionary change, complementing natural selection – to understand it and, if possible, find a way of expressing it in the form of a mathematical relationship.

It is well known that systems far from their equilibrium state show a propensity to form themselves into patterns or structures of increasingly complex order. To generate and maintain such order they need the driving force of some form of continuing input of energy, such as solar radiation, and are therefore referred to as dissipative structures – sustained by their continuing dissipation of the energy input. The theory of such production of ordered forms has been pioneered by Ilya Prigogine who describes these structures as islands of order in a sea of disorder.[269] Fritjof Capra writes of this development of order:

> The turbulent flows of water and air, while appearing chaotic, are really highly organized, exhibiting complex patterns of vortices dividing and subdividing again and again at smaller and smaller scales. In living systems, the order arising from non-equilibrium is far more evident, being manifest in the richness, diversity, and beauty of life all around us. Throughout the living world, chaos is transformed into order.[270]

Self-organization in the natural world is an idea that embraces the notion of self-making (*autopoiesis*) in which less complex structures form the building blocks of more complex structures. This implies that the intricacies of nature are not to be thought of as depending on the random assembling of mere molecules. It is important to note, too, that the term *chaos* does not imply total randomness or meaninglessness in the unfolding process of a physical system. It refers to a special sort of randomness within a limited range – the range of possible states or arrangements of the system, known technically as its *strange attractor*.

In its chaos mode of operation, a system's passage through successive states is ultra-sensitive to any small perturbations that might be introduced into it. If the system is completely isolated, its dynamical behaviour is claimed to be deterministic in that it follows precise mathematical laws.[271] Nevertheless, any uncertainty in its condition at any moment, or any minute disturbance from outside, is enormously amplified as its behaviour unfolds, so that the state of the system a little way into the future is unknowable – perhaps inherently so. This fundamental feature of exquisite sensitivity in chaotic systems such as the earth's atmosphere, is dramatically illustrated in the often-quoted idea of a butterfly flapping its wings and thereby ultimately influencing the behaviour of a hurricane on the other side of the world. It is known as the 'butterfly effect'.

It seems, then, that there is indeterminacy in both the quantum world of particle phenomena and the macroscopic world of physical systems; a certain degree of openness to the future is characteristic of both. But whether this is merely an *epistemological* openness (an indeterminacy in what can be known) or, alternatively, a genuinely *ontological* feature (an inherent property of physical reality in its very being) is itself an open question.[272]

Since the discovery of the structure of the DNA molecule in 1953 – hailed as the greatest advance in biology since Darwin's theory of evolution – some molecular biologists have been inclined to emphasise a reductionist view of a human being (or animal) as nothing but a survival machine, programmed to preserve the genes.[273] The emphasis here is more on the structural content than on the pattern or network of relationships, both internal and external:

> Whereas cells were regarded as the basic building blocks of living organisms during the nineteenth century, the attention shifted from cells to molecules towards the middle of the 20th century, when geneticists began to explore the molecular structure of the gene. Advancing to ever smaller levels in their explorations of the phenomena of life, biologists found that the characteristics of all living organisms, from bacteria to humans, were encoded in their chromosomes in the same chemical substance, using the same code script. ... Biologists had discovered the alphabet of a truly

> universal language of life. This triumph of molecular biology resulted in the widespread belief that all biological functions can be explained in terms of molecular structures and mechanisms.[274]

Chaos theorists on the other hand have tended to contradict that belief.

> The first chaos theorists . . . shared certain sensibilities. They had an eye for pattern, especially pattern that appeared on different scales at the same time. They had a taste for randomness and complexity, for jagged edges and sudden leaps. Believers in chaos . . . speculate about determinism and free will, about evolution, about the nature of consciousness and intelligence. They feel that they are turning back a trend in science toward reductionism, the analysis of systems in terms of their constituent parts: quarks, chromosomes, or neurons. They believe that they are looking for the whole.[275]

The reductionist approach has been extremely successful and continues to play an essential role in the search for understanding, but it does not enable one to deal with the fact that systems generally exhibit properties that are not readily explained in terms of their constituent parts. As often stated, the whole is more than the sum of its parts.

A 'THEORY OF EVERYTHING'

There is a frequent urge in science to search for unity of knowledge, to embrace the reductionist and the holistic and to make connections between seemingly disparate phenomena. For theoretical physicists the grail is an overarching description of the operation of all four of the fundamental forces – a unified theory often referred to colloquially as a 'Theory of Everything'. Already there are 'Grand Unified Theories' on offer which connect the *strong*, *weak* and *electromagnetic* forces but do not encompass the very different *gravitational* force. There is optimism, however, that the new so-called *Superstring Theory* will prove itself as a successful 'Theory of Everything'.[276] What would this mean? Will such a theory signal the end of theoretical physics?

A century ago there was overconfident talk that no more than a clearing up operation was needed in physics. In 1988 Cambridge mathematician Stephen Hawking wrote in similar vein: 'I still believe there are grounds for cautious optimism that we may now be near the end of the search for the ultimate laws of nature'.[277] However, the majority of physicists would no doubt regard this as a very bold supposition, bearing in mind the repeated discoveries this century of structures within structures in the world of particles. Besides, a unified theory of particle behaviour would not be able to provide any clue to the laws of nature operating at the higher levels of complexity. These higher level laws may indeed be constrained to some extent by the nature and behaviour of particles, but they emerge as genuinely novel, underivable from the knowledge of particle physics.

The search for a genuine 'theory of everything' – one that seeks to take into account not only the multi-levelled world picture of science but also questions about the meaning and purpose of the world and its living things – clearly leads us beyond science into the realms of metaphysics and theology. This quest is one of the main concerns of the revived natural theology that is emerging as part of Christianity's continuing engagement with the modern world. The quest is one aspect – a major aspect perhaps – of the broad range of discourse between theology and science. It takes place within a postmodern respect for differences of world-view but with a desire to know and understand the true nature of things as deeply as possible, giving full weight to basic Christian doctrine, however reconsidered, and to the well established findings of science.

Before attempting to provide a framework for a theological 'theory of everything' or meta-narrative, we turn to some of the main characteristics of the world-picture revealed by the sciences, especially physics.

FEATURES OF THE EVOLVING UNIVERSE

One of the great discoveries of the 20th century is that the universe itself has a history – a continuing process of expansion and evolution. We see a universe that is 'large and old, dark and cold', presumably containing a vast number of planets,[278] on at least one of which an astonishing variety of life has emerged, including human beings who can think and reflect upon the nature and meaning of it all. If our expanding universe is to contain observers like us (or, indeed, any carbon-based life-forms), it has to be old and correspondingly large – old enough to have passed through at least the life cycle of the first generation of stars and reached well into the life cycle of the second generation. It was from many of the earlier stars that the chemical elements emerged that were to be incorporated into the later planets, some of which became potential sites for the emergence of life.

> On at least one planet, and perhaps on millions, conditions of temperature, chemical environment, radiation, and the chance congregation of simple atoms, permitted the coming into being of quite elaborate molecules with the power of replicating themselves in that environment. In a remarkable interplay of contingent chance (to get things going) and lawful necessity (to keep them going) there had begun a process by which systems of ever-increasing complexity would evolve. On our planet this eventually led to you and me.[279]

In the beginning was the Big Bang, say the cosmologists, and everything in the universe – every event, every particle and movement, from the most powerful burst of gamma rays to the song of a bird – derives its physical energy from that initial moment. Ours is the first generation to possess such a comprehensive overview of cosmic history and, indeed, to be able to reflect on the fact that all the non-hydrogen atoms in the human body originated in the hot interiors of stars. At this advanced stage of an immense process we and all life-forms are being made from the ashes of stars.[280]

Four features of this new world-picture seem especially significant: firstly, the non-trivial fact that the properties of the universe are such as to allow the emergence of biological life (the Anthropic Principle); secondly, the interplay between novelty-producing 'chance' and law-maintaining 'necessity'; thirdly, the degree of correspondence between mathematical theories and the physical reality they describe, and fourthly, the emergence of a broad spectrum of levels of complexity. We consider each of these in turn.

THE ANTHROPIC PRINCIPLE

Reflection on the new cosmic history raises a question which especially interested Einstein: Does the universe have to be the way it is?[281] Is this the only possible way or could it have been differently arranged? This is a question that also intrigued Brandon Carter and Robert Dicke in the late 1960s when they noted several remarkable 'coincidences' amongst the huge numbers (as large as 10^{40}) associated with the ratios of certain constants of nature. Do these seeming coincidences imply a very special cosmos, they wondered, with just the right properties for its long fruitful development?

As they considered the fact that we exist in the universe at this particular stage of its history on this particular planet, they concluded that the time and place were not insignificant details. They drew attention to the fact that the existence of observers in it implies that the nature of the universe must be such as to allow their appearance on the scene at some time and place. This is a statement of the so-called Anthropic Principle in its weak version.[282] It is a statement that may seem self evident and therefore rather pointless, but there is the underlying implication that not any universe will allow the emergence of life. It seems that if the universe is indeed to be life-bearing it requires a very particular character, set by its laws of operation, fundamental constants, and initial conditions.[283]

The Weak Anthropic Principle is essentially a selection principle. Suppose there existed a vast range of universes – whether in reality or only in the mind of a cosmologist – each with its particular set of physical constants and starting conditions which determined how it develops. Our universe would be one of probably very few that would have the special materials and environments needed for life. Most would turn out to be barren. For example, if in our universe the gravitational force had been somewhat stronger, stars would have burned too quickly and the cosmic expansion might have been too short-lived – and with weaker gravity the primordial gaseous material would not have been drawn together sufficiently to form fiery stars, given the dispersive effect of the expansion. Again, if the strong nuclear force was two per cent stronger the di-proton could readily form and would allow stars to burn hydrogen into helium at a catastrophically fast rate – and if this force were five per cent weaker the deuteron (proton plus neutron), with its vital role of keeping the stellar burning process slow and steady, could not exist. Yet again, the strength of the weak nuclear force seems delicately poised to allow the vast flux of neutrinos from the interior of a large star not only to escape but also to help blow the star open as a supernova. As for the relative strengths of the strong nuclear force and the electromagnetic force, it is quite likely that the structure of the water molecule, with its key role in the sustaining of life, depends sensitively on their ratio. Any change of its layout would probably alter its capacity to play that role.

It seems, then, that the complex structures of biological life have come into existence through an extraordinarily intricate set of conditions and processes. A delicate balance is needed not only in the appropriate evolution of the cosmos as a whole, but all down the line in the evolution of planets and their lifeforms. Indeed, the universe seems remarkably fine-tuned to produce life.[284] Perhaps theoretical physicist Paul Davies speaks for many scientists when he states that, although not attached to any conventional religion, he has come to believe that the physical universe is put together with an ingenuity so astonishing that he cannot accept it merely as brute fact.[285]

CHANCE AND NECESSITY

On all scales of size and all levels of complexity the universe operates according to physical laws (which are not causal agents but simply statements of its regularities), showing consistency of behaviour while at the same time allowing the emergence of new forms. There is an orderliness in the entire process that is neither simple nor wildly complex but sufficiently complex to make for rich variety. The nature of the process is neatly summed up in the title of biologist Jaques Monod's celebrated book, *Chance and Necessity*. The complex evolution of living things has depended on the interplay of novelty-producing chance and life-sustaining necessity. Monod pictured life beginning by the chance assembling of simple molecules into macro-molecules capable of replicating themselves. These then reproduced themselves via chemical reactions which provided the preserving role of necessity. Again, in the neo-Darwinian account of evolutionary biology, genetic mutations (chance) are selected by competition and passed on in a relatively reliable way within a stable environment (necessity).[286] Interpreting the scene pessimistically, Monod writes:

> Pure chance, absolutely free but blind, (is) at the very root of the stupendous edifice of evolution .... The ancient covenant is in pieces; man at last knows that he is alone in the unfeeling immensity of the universe, out of which he emerged by chance.[287]

Others have seen it in exactly the opposite way. As Polkinghorne puts it:

> Chance is the engine of novelty in an evolutionary world; necessity is the selector and guardian of fruitfulness. A world of pure chance would be chaotically infertile; a world of pure necessity would be rigidly infertile. It is the interplay of the two tendencies that produces the actual fruitfulness of our universe.[288]
>
> When I read Monod's book I was greatly excited by the scientific picture it presented. Instead of seeing the role of chance as an indication of the purposelessness and futility of the world, I was deeply moved by the thought of the astonishing fruitfulness that it revealed inherent in the laws of atomic physics ... (T)he action of chance is to explore and realise that inherent fruitfulness.[289]

Peacocke points out that Monod gives undue metaphysical status to the randomness of events at the molecular level – calling it 'blind chance' – for it is precisely the randomness of genetic mutations that enables 'the full gamut of potentialities of living matter' to be explored. Peacocke sees chance not as blind but as if it were 'the search radar of God, sweeping through all the possible targets available to its probing'.[290]

THE UNREASONABLE EFFECTIVENESS OF MATHEMATICS

The scientific picture often evokes admiration about the ingenuity of it all. What is also remarkable is the ability of the human mind to formulate that picture in mathematical terms. In fact, between pure mathematics and the basic structure of the material world there is an intimate connection that is usually taken for granted and yet is quite surprising.[291] The theoretical physicist Eugene Wigner wrote a paper entitled 'The Unreasonable Effectiveness of Mathematics in the Natural Sciences' in which he remarked that the miracle of the appropriateness of the language of mathematics for the formulation of the laws of physics is a wonderful gift that we neither understand nor deserve.[292] Parts of mathematics have arisen from explorations of uncharted areas of abstractness, guided by a desire for conceptual beauty and unhindered by any concern for relatedness to the physical world. Some of these abstract formulations have in due course proved to be just right for modelling aspects of nature. For example, Einstein found that an existing geometry of curved space was just what he needed to construct his General Relativity and, some decades later, a particular algebra of symmetry groups turned out to be just right to account for the family groupings of elementary particles.

THE HIERARCHY OF LEVELS OF COMPLEXITY

With this remarkable mental capacity for mathematics and other abstract reasoning, human beings have come to recognize their deep rootedness in a vast process of cosmic and biological evolution and to know that their particular place in nature has required the full range of space, time and multi-levelled complexity of the material universe.

We ourselves are constituted as a hierarchy of levels of structure (molecules, cells, organs, brain/mind) and the way in which each level operates depends on the workings of both higher and lower levels – the properties of organic cells are not derivable from the properties of molecules but they are to some extent constrained by the latter, and what the molecules actually do in the working of a cell is set by the overall task of the cell itself. There is bottom-up and top-down causation or interaction between levels, applicable at all levels it seems.

In each level the entities are made up of the items of the level immediately below, having emerged from it in the evolutionary process. It is important to note that the specific behaviours, relations and properties at any given level are not reducible to the features of the constituent elements – the concepts of cell biology for example are not reducible to the concepts of molecular physics. In other words, the laws and regularities of cell biology are not predictable from those of molecular physics – and this applies all the way down the complexity scale, from molecules to atoms to nucleons/nuclei to quarks/leptons, and surely to the hypothetical particles of superstring theory, as well as up the scale from cells to organs to complete life forms.

From flora and fauna there are the yet higher levels associated with animal populations, ecosystems and the earth's entire biosphere (to say nothing of the physically larger but simpler assemblies within the cosmos); and at the most complex we have the realm of the human sciences – psychology, the social sciences, ethics, etc.[293] Even the natural sciences on their own can give rise to questions of a metaphysical nature – Why does the universe exist? Why does it operate in regular lawlike fashion? Why are the laws of nature as they are? – but when the human concerns of value and meaning, of truth, goodness and beauty, enter the picture, the metaphysics readily takes on a theological complexion.

If the development of this richly textured universe is indeed a finely-tuned open-ended evolutionary process that is deeply accessible to human understanding, what are the implications for Christian theology? And can a renewed theological understanding of the world help to place the endeavours of science in a wider context of meaning and value?

5 THE PRESENT SCIENCE-THEOLOGY DISCOURSE

THE CHALLENGE OF NEW KNOWLEDGE

Before considering the present renaissance of the discourse between science and theology,[294] let us note that in the late Middle Ages a change occurred in the relationship between the two disciplines. With the establishing of the institution of the university in early 13th century Europe, Aristotle's scientific writings became available and began to make a strong impact on theological thought. Until then, natural philosophy had been regarded as subservient to theology, useful for the aid it could provide in interpreting Holy Scripture. Then:

> For the first time in the history of Latin Christendom, a comprehensive body of secular learning, rich in metaphysics, methodology, and reasoned argumentation, posed a threat to

> theology and its traditional interpretations ... (T)he Aristotelian world system was not readily reducible to the status of a theological handmaiden.[295]

There was a growing uneasiness among traditionally minded theologians, for whom parts of Greek philosophy were dangerously subversive, severely constraining the doctrine of God's omnipotence.[296] Opposing the traditionalists were those more liberal theologians who found Aristotle's natural philosophy essential to a proper understanding of God and his creation.[297] This was the background against which Thomas Aquinas worked to bridge the divide between traditional theology and the new Aristotelian world-view.

The context of today's theology is somewhat similar. As in the late medieval period, there is a certain tension between the conservative and liberal theological positions concerning the degree to which traditional Christian doctrine needs to be attuned to the huge body of new knowledge that has been accumulated over the past four centuries. Those on the conservative side tend to work within one of the older paradigms and have yet to engage seriously with the science-theology discourse.[298] Another group who have not entered the discussion are those orientated towards political and liberation movements, seeing the philosophical orientation of Science & Religion as a distraction from responsibility for the poor and oppressed.[299] So far it has been those scientist-theologians and systematic theologians of a less conservative persuasion who have taken the debate seriously, with viewpoints ranging from those of traditional Trinitarian doctrine to a strongly naturalistic empirical theism.[300]

Table 3 (on the next page) shows one of the simpler ways of summarizing the wide spectrum of positions in the interaction between science and theology.[301] Apart from the monistic views of scientism and creationism at the extremes of the spectrum (in which the world is interpreted entirely in terms of a single overriding principle) and the dualistic perspectives of science-and-spirituality and faith-and-science, the table shows a broad central stance in which there is active engagement between the disciplines. Here, in specialist gatherings of scientist-theologians, whose background is primarily scientific, and other scholars who are primarily theologians, a strong consensus has arisen about the evolutionary nature of the world in both its cosmic and biological aspects.[302] On the other hand, differences amongst scientist-theologians have emerged about the nature of divine action in the world, and their opinions diverge yet more strongly over the question of the degree to which science and theology can be integrated and, consequently, over ontological questions about the person of Jesus Christ.[303]

These differences reflect the greater complexity and abstractness of ideas as one moves across the disciplinary spectrum from the natural sciences, through the social sciences and history, to theology. On the other hand there is a fair degree of consensus within the disciplines of science and theology concerning the need to resist postmodern scepticism concerning the possibility of any knowledge of reality.

Table 3. Spectrum of Relationships between Science and Religious Belief

non-theistic science	**science & spirituality unlinked**	**science & theology interacting**	**faith & science unlinked**	**non-scientific theism**
for example: Monod Dawkins Atkins	many religious scientists	for example: Barbour/Peacocke/Polkinghorne[1] Whitehead/Cobb/Griffin[2] Torrance/Pannenberg/ Nebelsick[3]	many religious believers	for example: Bryan Gish Morris
[1]*scientist-theologians*, [2]*process philosophers/ theologians*, [3]*systematic theologians*				

Generally speaking, scientist-theologians regard these disciplines as providing together the data and insights needed for a cogent unified account of the world – a world that is assumed to be real and the bearer of meaning. They tend to hold to a philosophical view of the nature of knowledge known as *critical realism*, one of the three epistemic approaches described in the following section.

EPISTEMIC ASPECTS

The philosophical movement of *logical positivism* (or *logical empiricism*) became fashionable in Western thought during the early part of the 20th century. Following Kant, logical positivists distinguished between analytic statements (those which are necessary and certain simply because they are tautologous, such as 'All bachelors are unmarried') and synthetic statements (those expressing something empirical about the world, such as 'The sun sets at ten past six'). The latter are statements capable of verification via some sense experience, at least in principle. All statements not falling within these two groups were regarded as meaningless, incapable of being either true or false – and this included much of metaphysics, ethics and theology.[304]

AJ (Sir Alfred) Ayer's *Language, Truth and Logic* (1936) was a classic statement of logical positivism in its most aggressive and iconoclastic phase.[305] At that time Ayer was able to dispense with the existence of God in a few pages. He was not even an atheist, he said. How could one deny the existence of God if one did not even know what the word 'God' meant? Fifty years later, when asked what he regarded as the main defects of logical positivism he replied 'Well, I suppose the most important of the defects was that nearly all of it was false'.[306] He still held to a weak form of the verification principle in which (rather vaguely) the appeal was made to some sort of ideal observer to do the verifying. He added that it is possible that the existence of God can be claimed as an explanatory hypothesis, but that in his own experience it explained nothing.[307]

The assertions of logical positivism continue to ripple through popular consciousness and it is not unusual to hear of the existence of God discussed as a matter that needs firm scientific verification, sought as a prerequisite to religious belief. However, the analytic/synthetic distinction and the verification principle have been rigorously challenged and found wanting.[308] Furthermore this once fashionable epistemic view sidelines much of the metaphysics, ethics and theology that constitute domains of human life and thought that seem central to our understanding of what it means to be human. Consequently it has very largely faded from the philosophical scene that it has enriched, having brought a fresh awareness of the need for clarity and precision in the use of language – an insistence on explaining exactly what one means and on analyzing the meanings of terms used, especially a profound term like 'God'.

Perhaps the most widespread of epistemic approaches – certainly in scientific fields but also in the enterprise of systematic theology – is that of *critical realism.*[309] Here the subject of inquiry, whether it be the natural world or the being of God, is regarded not as simply a construct of the human mind but as actually existing independently of the mind – it is a *realist* view. Furthermore, since knowledge of the subject is acquired through the use of critical faculties, acting like spectacles through which the subject is viewed and reflected upon, this view is called *critical* realism. It stands in contrast to the *naive* realism of earlier centuries since it does not claim exactness of fit between mental models and the reality to which they refer.[310] It acknowledges that knowledge is provisional, corrigible, approximate – and, from a later point of view, frequently wrong.[311]

Despite these limitations and the scepticism of some postmodern critics concerning notions of objective reality (especially the idea of access to it), both science and theology continue to be seen by their critical realist practitioners as engaged in a quest for understanding through the task of constructing increasingly accurate maps of reality.

Generally speaking, scientists do not lay claim to absolute knowledge but rather to an increasing verisimilitude (similarity to what is actually the case) between their theoretical models and the reality they describe.[312] For example, the development of the understanding of the nature of light from the time of Newton to the present day has passed through several increasingly advanced phases. Its present description in terms of the theory known as quantum electrodynamics is a huge advance on Newton's notion of corpuscles of light. When this theory is used to calculate certain fundamental physical quantities, it yields values that are remarkably close to the measured values. Again, a much more profound and accurate picture of the structure of atoms has been acquired over the course of the 20th century.

Theologians, likewise, can look back to a pattern of revelation which has unfolded in human history,[313] even if with much less consensus than is usually achieved in science, for theology involves the exercise of subjective judgement and intuition to a far greater extent. For example the biblical account of the history of Israel shows a progressive accumulation of insight into the character and purposes of God – a development which, from the point of view of the Christian Church, continues through the New Testament writings and far beyond. Again, the early Christian understanding of the person of Jesus Christ was enriched through the reflections of successive generations of thinkers and writers, showing a progression from the Pauline letters to the Gospel of John and then to the sophisticated formulations of the Council of Chalcedon (in the year 451).[314]

The third of our approaches is that of *contextual coherence*, a detailed description of which has been given by Niels Gregersen.[315] Here the concern is for coherence between different beliefs and practices insofar as these are interconnected within a logically consistent and substantially comprehensive pattern of thought. As in critical realism there is, firstly, an insistence that it is some aspect of reality that is pursued (not simply a mental construct) and, secondly, an acceptance of the context relatedness of each part of knowledge, much of it culture dependent. But unlike critical realism, contextual coherence theory harbours a certain scepticism toward the claim of increasing verisimilitude. As Gregersen writes:

> coherence theory ... does not have to insist on the claim that human knowledge, in general, entails a steady 'approximation' to reality... (It) also rejects the possibility that isolated 'propositions' can be shown to be true by 'corresponding to facts', mundane or divine.[316]

Because the 'facts' of reality itself are inaccessible, the only workable criterion of the truth of a theory (for the coherentist) is that of its internal coherence and its being in accordance with a wider pattern of statements that are considered to be true.

Unlike a foundationalist view in which knowledge is based on a few allegedly indubitable facts or propositions and is built up in the form of a chain of successive arguments, the coherence model expresses a view of knowledge as analogous to a raft of different planks.[317] The construction of the raft begins with the assembly of a wide range of planks – data and knowledge, scientific and non-scientific – and only then are the implausible elements ruled out. The search is for unity of knowledge through piecemeal construction, drawing upon data and theories from various contexts – on the one hand the basic data of human experience and on the other hand theories that emerge from various levels of abstract thought as the best available truth-candidates.

Gregersen suggests that the strength of contextual coherence theory is that it allows for a plurality of epistemic theories and visions within a common framework of rationality. It is thus in tune with the postmodern affirmation of the diversity of paradigms and frameworks of knowledge but, more than that, it is especially orientated to the task of raft-building – to the creating of some holistic metaphysical multi-faceted scheme provided this does justice to the inherent richness of reality.

Critical realists and coherentists alike are concerned with the question of the assessment of ideas and theories, looking always for consonance and coherence between the mode of knowing and the reality known.

Polkinghorne often stresses the point made by theologian Thomas Torrance that components of reality (from electromagnetic fields to the being of God) are known in ways that accord with their natures and that we cannot determine beforehand what these epistemological modes will be. Torrance likes the metaphor of listening, of being receptive to the impact of reality as it comes to us as a gift.[318] He urges that in our search for scientific knowledge we need to

> allow fresh revelations of nature to call forth from our minds new modes of thought by which the process of our minds can be brought into the most complete harmony with the process of nature.[319]

Theologians too must operate

> in accordance with the fluid axiomatic modes of thought that have served so well to reveal the mysteries of nature ... Theological science must therefore generate concepts of God that are *worthy* of him ... concepts that are forced on our minds by the sheer nature of the divine Majesty.[320]

Such similarities of approach are not always appreciated. In the popular stereotype, writes Ian Barbour, the scientist's theories are tentative hypotheses that are continually criticized and revised, while religious beliefs are unchanging dogmas that the faithful accept without question. In fact *both* disciplines engage in the continuing activity of corporate assessment in the quest for consonance between reality and its conceptualization. Within each of the various natural sciences there is a thorough and continuing process of intersubjective evaluation of theories which is one of the hallmarks of science. Likewise, within the community of Christian belief and praxis, theologians go through a largely similar process of intersubjective evaluation as they reflect on the pattern of divine revelation, even if with much less consensus than in science.

In assessing new hypotheses and theories, scientists value especially the qualities of economy, coherence and explanatory and predictive power, and they are often strongly influenced by the aesthetic aspect – the beauty and simplicity of an underlying idea or the elegance of a mathematical formulation. All these factors contribute to the conviction that the theory under consideration has about it the ring of truth. Despite the vastly different subject matter, there is in theology too a quest for a description and explanation that is coherent, matches experience well and is widely explanatory. Thus in both fields, knowledge requires acts of judgement and commitment involving all the antennae of reasoning and intuition. In that sense it is personal.

This personal nature of knowledge has been explored in depth by the scientist-philosopher Michael Polanyi (1891-1976), especially in his oft-quoted book *Personal Knowledge* (1958). To a large extent he was ignored by the philosophers of Oxford when he retired to that city from the University of Manchester, yet some of his ideas seem especially relevant to the science-theology discourse that has developed since then.[321]

In emphasising the fundamental unity of *knower* and *known* over the whole range of knowledge, Polanyi stood sharply against logical positivism and its ideal of pure impersonal objectivity, purged of the interpretive element that enables the knower to attribute meaning and value to the subject matter.[322] Furthermore, he makes the vital suggestion that whatever has more depth of meaning, and is thus more attractive to our minds, is more real.[323] This implies that if the material world is to be regarded as objectively real then, *a fortiori*, so too is the realm of moral values – a necessary assumption for the construction of any metaphysical linking of science and theology (to be discussed later).

One of Polanyi's key epistemic ideas is the notion of 'tacit knowledge' – that which is an accumulation of well-winnowed taken-for-granted wisdom within a scientific community or any other cultural group. It is knowledge that cannot altogether be articulated. As Polanyi puts it succinctly, we know more than we can tell. When theologians assess a doctrine of divine creation or a particular religious experience or practice, or when scientists assess a hypothesis of biological evolution or the results of a scientific experiment, they look at the matter through the spectacles of their tacit knowledge, using the latter as a diagnostic tool (often unconsciously) to feel their way to a deeper understanding of the situation.

Another basic idea in Polanyi's epistemology, drawn from many years reflection on the dynamics of scientific discovery, is that of the power of 'tacit inference', a term he uses largely as a synonym for intuition. This is the mental faculty which, in concert with creative imagination, comes into play in the quest for pattern and meaning in the data of experience or experiment. The dynamism in scientific discovery, writes Joan Crewdson,

> is due in part to the deliberate effort or thrust of the imagination searching for clues, in part to the spontaneous activity of intuition which integrates what imagination has hit upon, and in part to the inexhaustible nature of the hidden reality. Polanyi's analysis shows imagination serving the questing mind much as the blind man's stick probes the environment, searching for new coherences and guided by intuition towards a possible solution.[324]

Altogether Polanyi presents a picture of a richly textured process of the building of human knowledge and understanding – one that is far more than an application of rules of logic or of objective criteria, such as verifiability or falsifiability.[325]

The concept of tacit inference in scientific research resembles what John Macquarrie calls 'architectonic' reasoning – the necessarily speculative and imaginative element required in the construction of a metaphysical or theological fabric (such as Karl Barth's *Dogmatics*).

> I have in mind a constructive use of reason in which we build up rational wholes, theories, or interlocking systems of ideas, but do so not by deductive argument but rather by imaginative leaps which, so to speak, integrate the fragmentary elements in inclusive wholes. Of course, such imaginative leaps need to be immediately tested and

> subjected to scrutiny, yet something of this kind goes on in most intellectual disciplines, including the natural sciences.[326]

He adds that this architectonic function of reason would seem to have something in common with aesthetic sensibility, a point which we meet later.

For both critical realists and coherentists a fundamental aspect of epistemic development is the question of making a choice between alternative theories or models – a choice that may involve the criterion of 'inference to the best explanation'.[327] Where direct cogent evidence is scarce or not available and a theory is therefore necessarily based on indirect and circumstantial evidence – as in the case of the historical sciences such as cosmology and geology where the full process under consideration cannot be put to the test – the plea for its validity rests on this criterion. Here a theory or model is judged by the overall impression it makes in terms of the afore-mentioned criteria of economy, simplicity, coherence, comprehensiveness, and overall explanatory power. The more it shows these qualities the more will it resonate as a convincing explanation. And these are the qualities one would look for in any metaphysical account of the grand epic of the evolving cosmos.

Such epistemic questions continue to lie at the forefront of the science-theology debate as attention is given to aspects of God, humanity and the cosmos, and to the task of creating some sort of unified metaphysical account of the world and its meaning.

DIVINE TEMPORALITY

One of the present concerns in the debate is the question of how God relates to the passage of time – what this implies for the existence of human free will and, indeed, the freedom of the physical world to unfold strictly in accordance with its divinely established laws. In the earlier section on complexity theory it was suggested that there is indeterminacy in this unfolding, not only in the epistemic sense but perhaps also in the very nature of physical reality. Any such ontological indeterminacy is important for the discussion of divine temporality and divine action.

Does God know the future in all its detail? In classical theism a divine vantage point is envisaged that is outside time, providing an atemporal view of a 'block universe' in which all of space-time is seen 'at once'.[328] On the face of it this seems appropriate for the God who is described as omnipotent, omniscient, unchanging, and impassible (not subject to suffering) – aware of all changes in the world without being subject to change himself. However, two objections have been raised. Firstly, such a picture does not seem compatible with the reality of the free will that we as humans instinctively feel we possess – and the freewill of human beings would surely be the Creator's intention in order that we may have the responsibility of making moral decisions. Secondly, it is argued that the idea of a block universe is also not compatible with what seems to be a genuine open-endedness in nature's processes. Our world of *being* and *becoming* unfolds moment by moment and has a real history and a real present moment, but its future moments, it is suggested, do not yet exist – they are not there to be foreknown, even by God. According to Polkinghorne,

> God must know (the universe) in its temporality, as it actually is. God must not just know that events are successive; they must be known in their succession. This implies that temporal experience must find a lodging within the divine nature.[329]

In similar vein Peacocke suggests that the idea of God as personal can only have meaning if God experiences something like the succession of conscious states that we ourselves experience as persons, since, humanly speaking, to be a person with consciousness *is* to be aware of a succession of states of mind. Like Polkinghorne, he does not accept the notion of divine foreknowledge. In any case, Peacocke envisages a divine self-limiting (of power and knowledge) that allows the natural order to be unconstrained in its process of exploration and development[330] – and any such *self*-limitation does not imply that God is less than omnipotent or less than all-knowing with respect to that which can be known.[331]

In process theology[332] the nature of God is conceived of as dipolar, both *temporal* and *atemporal*.[333] This is widely accepted in other theological thought, too, since it fits well with the complementary aspects of the *being* and *becoming* of the God who is the faithful fount of love in eternity and a continuing expression of love in history.[334] The temporal pole is seen to be an essential aspect of what it is to be *personal* and thence to be aware of a succession of states of mind.[335] Thus the notion of openness in physical process is linked to the questions of divine temporality and human free will – and ultimately to the metaphysical question of the kind of universe that might be expected as the creation of a personal God of utmost love, who responds to prayer and warns through the prophets.

DIVINE ACTION

A longstanding topic in the debate is that of divine action. If the world is no longer construed in terms of the mechanistic Newtonian picture but rather as a world of flexibility and openness to novelty, what is the manner and scope of divine action and wherein lies the causal joint? Where does God actually *act*?

The enterprise of *process theology* offers a starting point. It is based on the *process philosophy* of Alfred North Whitehead which is perhaps the most important attempt in the 20th century to form a new systematic metaphysics that unites science and religion.[336] Whitehead maintained that religion contributes its own independent evidence which metaphysics must take into account. Claiming that the concepts of religion are of universal validity, even if they are derived from special experiences and events, he writes:

> The dogmas of religion are the attempts to formulate in precise terms the truths disclosed in the religious experience of mankind. In exactly the same way the dogmas of physical science are the attempts to formulate in precise terms the truths disclosed in the sense perception of mankind.[337]

Process theology is more concerned about *becoming* than about *being*. It emphasises the open-ended process of the world as an ever developing series of interdependent but discrete events. Each event is regarded as an entity in its own right with its own individuality, a centre of spontaneity and self-creation, forming part of an interacting network of individual *moments of experience*.[338] Each is characterized by a *prehensive* phase in which there is, so to speak, a menu of possibilities for that event, emerging from past happenings, and the divine role is to provide a lure towards the choice of the good, without coercion. This is followed by a *concrescent* phase in which one of the possibilities is actualized.

Process theology places a strong emphasis on the immanence of God in the world – the Divine Persuader who gently seeks to work with the recalcitrant material of the world. This is not the omnipotent, omniscient, unchanging God of classical theism but One who, through immersion in the process of the world, not only works in it everywhere but is affected by it and in that way develops with it. *Creatio ex nihilo* is not a notion that is readily accepted in process theology. God's role is one of *creatio continua* with the eternally existing material of the world, ever urging it towards the three qualities of peace, truth and beauty. The stress on divine immanence suggests a *panentheistic* view of the world – one in which all that is created exists *in God* – even if the being of God is thought of as dipolar, both *in* time and *beyond* time.[339]

Reflecting on how the question of divine action may be understood in the scheme of process theology, Polkinghorne finds two major difficulties. One is the jerky discreteness assumed in the world's process.

> I do not see that the physical world, as disclosed to scientific exploration, can be held to correspond to a concatenation of events in the manner suggested. Quantum physics involves both continuous development (described by the Schrödinger Equation) and occasional sharp discontinuities (measurements of quantum events) but it does not suggest the discrete "graininess" that process thinking seems to suppose.[339]

His second difficulty is theological. Whitehead's metaphysics, he remarks, assigns a highly attenuated role for God – a role of persuasion in the entire process that lies too much at the margins of the world and falls far short of the providential care and the fulfilment of ultimate hope envisaged for the God of Christian tradition. Neither the role of *creatio continua* alone (as in process theology) nor the initiating but limited act of *creatio ex nihilo* (as in deism) does justice to the nature and purposes of God as understood from centuries of Christian reflection.

In response to such reflection, one strategy is that of Austin Farrer's suggestion of 'double agency', invoking a divine primary causality that is in some way present, but hidden within the secondary causalities described by science. He wrote of God's working omnipotently on, in and through creaturely agencies, without either forcing them or competing with them – a suggestion criticised by Polkinghorne as so mysterious that there is little one can say about it, and mentioned by Peacocke as a paradoxical notion that comes close to mere assertion of its truth.[341]

Of course, if a stronger primary causality is at work within *everything* that happens, this raises acutely the problems of evil and suffering in the world. Can divine action be conceived in a manner that takes due account of these problems while also drawing upon the insights of the trinitarian doctrine of God and the scientific picture of the world? Distancing himself from the panentheism often expressed in the works of Barbour and Peacocke, Polkinghorne writes:

> Many of us would share a recognition of the need to correct classical theism's undue emphasis on the transcendent remoteness of God, without feeling that this implied a necessity to adopt panentheistic language. It simply requires a recovery of the balancing orthodox concept of divine immanence.[342]

What is also needed is that the world be non-deterministic – not subject to the tightly bound sequence of physical cause and effect imagined by Pierre-Simon Laplace, but rather characterized by genuinely inherent indeterminacy in the workings of the material world, whether at the level of particle behaviour (according to quantum theory) or at the level of macroscopic systems (according to chaos theory), or both. There need to be 'causal gaps' if there is to be room for divine manoeuvre, as opposed to the 'explanatory gaps' that have at times been invoked in urgent attempts to make philosophical room for God. Furthermore, it seems important to avoid any idea of divine *intervention* into the operation of physical law, since the latter is deemed to be created and sustained by One who is the faithful continuing ground of nature's workings, not the author of occasional capricious forays into nature.

Ontological indeterminacy is commonly assumed by physicists in the case of the quantum world even though an ingenious deterministic interpretation (by David Bohm) is on offer. At this stage there is far less consensus in the case of chaos theory. Does the unknowability of the unfolding behaviour of a chaotic system imply that it is actually (ontologically) indeterminate, completely open to all the possibilities available to it within the range of its strange attractor? Polkinghorne is prepared to affirm macroscopic indeterminacy as a useful metaphysical assumption on which to form a hypothesis of divine action. Nancey Murphy and others do not accept this. They prefer to start from the level of particles since they regard the universe as highly deterministic at its macroscopic levels.[343]

If quantum indeterminacy is chosen as the place of bottom-up divine action, the latter can be envisaged in a minimalist or more widely embracing way. Either it is thought to be limited to the firing of neurons in the human brain in order to provide some sort of communication from God, giving visions of the good, or it is imagined as a route to the exquisitely sensitive macroscopic levels of nature. However, the way in which the microworld of quantum events connects to the macroworld of material process is not understood. The appeal of the minimalist's limiting of divine action to encouragement and illumination via the neural realm is that it sidesteps the theodicy question – how to reconcile the alleged omnipotence of a compassionate God with the existence of evil and suffering. The stronger one's account of divine action the more perplexing must become the tragedies of life.[344]

If on the other hand the way of divine action is considered to be directly through the subtle and supple behaviour of macroscopic process, this allows a comparison to be made with our experience of human agency.[345] Here there is a psychosomatic link between mind and body through which a person's physical action is initiated. The action can be thought of as one of a large array of possible physical movements, constrained by the nature of bodily structure and constituents – but it can also be considered from the holistic point of view of the person's decision. As mentioned earlier, there is both bottom-up and top-down causation at all levels of physically constituted hierarchies.

The concept of top-down causation is invoked by both Peacocke and Polkinghorne, but in different ways. Peacocke speaks of the relationship between Creator and creation in panentheistic terms, placing great emphasis on the immanence of God who is all the time creating in and through the processes of the world. These processes, he suggests, are themselves God's action and are constrained to be what they are in all their astonishing subtlety and fecundity by virtue of the way God interacts with the world-as-a-whole. Knowing the interconnectedness of the world to the finest detail, God is envisaged as being able to interact with the world 'at a supervenient level of totality' – holistically – thereby bringing about particular events and patterns of events.[346] Such interaction amounts to the input of information of a pattern-forming nature, the energy content of which can be vanishingly small so that there is no breach in the causal network of natural law. It is a form of top-down causation that Peacocke prefers to call 'whole-part influence' and it meets his concern always to interpret the world's happenings as naturalistically as possible, seeing this as a crucial task of theology in our scientific age.

Polkinghorne also speaks of top-down causality through the provision of similarly energy-less 'active information', but he suggests a more direct input into the world's processes. With the *chaos* concepts of 'butterfly effect' and 'strange attractor' in mind, it is conceivable that pattern-forming information can lead a system from one arrangement to another – since any trajectory from one point within its strange attractor to another does not involve any change of total energy – and Polkinghorne suggests that the divine will could thus be exerted within any macroscopic part of the world's structure. He believes, too, that there is a greater dynamical openness for divine agency via chaotic systems than simply through holistic operation on the world-as-a-whole. When challenged that macroscopic physical systems, even in their chaotic mode, follow deterministic equations and therefore cannot be expected to offer any room for maneouvre, he replies that the equations can be understood as *approximations* to true physical reality, applicable in only those rare and specific situations in which a system can be treated as totally isolated from its environment.[347]

The idea of divine providential action through hiddenly introduced active information is consonant with that of a gracious Creator who allows the creation to be itself and to have room to develop through the exercise of human free will and the pathways of free process, via divinely installed guiding principles of chance and necessity. In Christian theology it is the

Creator-Spirit who is thus creatively at work throughout space-time.[348] This Spirit of Life, referred to by John Taylor as the Go-Between God,

> is ever at work in nature, in history and in human living, and wherever there is a flagging or corruption or self-destruction in God's handiwork, he is present to renew and energize and create again .... If we think of a Creator at all, we are to find him always on the inside of creation. And if God is really on the inside, we must find him in the process, not in the gaps. We know now that there are no gaps .... If the hand of God is to be recognised in his continuous creation, it must be found not in isolated intrusions, not in any gaps, but in the very process itself.[349]

Arthur Peacocke likens the role of the Creator to that of the composer

> who, beginning with an arrangement of notes in an apparently simple subject, elaborates and expands it into a fugue by a variety of devices of fragmentation, augmentation and reassociation .... Thus might the Creator be imagined to enable (the unfolding of) the potentialities of the universe which he himself has given it, nurturing by his redemptive and providential actions those that are to come to fruition in the community of free beings – an Improviser of unsurpassed ingenuity – a composer extemporizing a fugue on a given theme.[350]

While the exercise of freewill seems to be a readily acknowledged fact of human life, the openness of physical process is a less common idea and so, time and again, when a natural disaster occurs the question is asked, Why did God allow it? – could there not have been some compassionate intervention? But in the wisdom of God the universe has been given the gift of free process, it seems, for its immensely creative development.

> Austin Farrer once asked himself what was God's will in the Lisbon earthquake (*which killed tens of thousands of people on a Sunday morning in 1755*). His answer – hard but true – was that the elements of the earth's crust should act in accordance with their nature. God wills neither the act of a murderer nor the incidence of cancer, but he allows both to happen in a world which he has granted the freedom to be itself. (*Italics added*).

If the way of the Creator with the world is to allow it freedom to be itself, not intervening intermittently, the tightly knit nature of the world makes it probably inevitable that what emerges is not just beauty and goodness but also natural disasters and the expression of evil. The biochemical processes that enable cells to mutate and make evolution possible are also those that produce cancers. Besides, it may well be the case that, given the intimate link between mind and brain, the free exercise of the human mind can only take place if the Divine Composer allows a high degree of freedom in nature's working out of its potentialities.[352]

Theology has to deal with a picture of the world which can be interpreted in a purely naturalistic way, yet allows room for the immanent Creator-Spirit to work alongside human agency and the ordinary processes of nature. The hiddenness of divine action will mean that it will not normally be distinguishable from natural process. But perhaps at a more profound level of meaning, divine action will be discernible to the eye of faith – seen on the one hand as creating and sustaining nature in all its regularities, and on the other hand operating through special top-down inputs of information, always in accordance with divine faithfulness and graciousness, seeking to create goodness and beauty through the precarious route of free development and choice.

DIVINE SELF-LIMITATION (KENOSIS [353])

Discussion of divine action is linked to the question of the purposes of God and suggests the setting up of a braod theological model of the creation. Jürgen Moltmann describes various models of the God/world relation and, as a theological 'thought experiment', looks for one which best relates to the scientific picture of complex evolutionary systems in nature.[354] In the model of God's handiwork in traditional Christian theism, God creates, maintains, saves, and perfects the world. Moltmann commends the elaboration of this model in trinitarian terms, introducing the idea of God's self-limitation (what the Jewish Kabbala calls *zimsum*, meaning concentration or contraction) – a self-shrinkage, so to speak, that allows ontological space for that which is created.[355]

> Must we not say that this 'Creation outside God' exists simultaneously *in God*, in the space which God has made for it in his omnipresence? Has God not therefore created the world 'in himself', giving it time *in* his eternity, finitude *in* his infinity, space *in* his omnipresence and freedom *in* his selfless love?[356]

Related to this is the powerful concept of the *kenosis* of God, a key idea on which to base a metaphysical theory of the 'epic of evolution'. Whereas classical theism's stress on the transcendence and omnipotence of God presents us with the problem of theodicy, the immanence of a kenotic self-giving God who suffers in and with the creation presents us with the question of divine vulnerability, taken up in W H Vanstone's frequently quoted ideas which we meet later.

So far we have considered critical-realist ideas of science and theology as elements in the search for understanding – a search which lies firstly in the cognitive realm. Introduction of the attribute of *kenosis* immediately opens the debate to the dimension of morality and its origin. Is the moral realm simply a construct of human communities – emerging through processes of cognitive evolution – or is it of divine origin, an aspect of objective reality that calls for a realist approach?

We turn first to a limited metaphysical scheme which represents a broadening of the scientific world-picture, bringing together the realms of the physical and the moral.

A THEISTIC COSMOLOGY

Cosmologist George Ellis holds a critical-realist stance *vis-a-vis* not only the physical universe but also the realm of moral values. His concern has been to seek a broadening of the scientific world picture into a 'cosmology' that includes the area of human values and aspirations.[357] His approach is to put the Anthropic Principle (concerning the fine-tunedness of our life-bearing universe) into the context of a Christian view of the highest good in human life and thence construct a model, or framework of understanding, of the nature and purpose of the creation.

Noting that the universe exists as a hierarchy of levels of complexity and meaning, Ellis points to the vivid human experience of moral action (in the face of adversity and sometimes danger) as a way of knowing. In particular he sets forth the quality of *kenosis* as the widely accepted heart of morality. Morality, he claims, is a more fundamental strand in the foundation of the universe than the physical. Physical reality cannot, from within itself, give rise to ethical concepts, but the ethical can lead to the physical if the two realms lie within the purpose of a designed universe. So, with Christian belief in mind, he takes as his starting point the axiomatic statement that

> *there is a transcendent God who is creator and sustainer of the universe, whose purpose in creation is to make possible high-level loving and sacrificial action by freely-acting self-conscious individuals*

and argues in detail that the universe is excellently matched to this fundamental assumption.[358]

As in the case of Darwin's theory, here again is an example of the hypothetico-deductive approach – letting the informed imagination create a hypothesis and then deducing particular consequences that can be tested by experiment or observation.

Apart from the ordered anthropic nature of the universe – favourable for the emergence of conscious beings with free will and equally provident towards all such beings, however morally good or bad – there are two more conditions necessary for high-level kenotic behaviour to occur freely: the hiddenness of God, which allows the freedom of response necessary for genuine moral choice, and the possibility of divine revelation, which provides knowledge and visions of ultimate reality to those open to them. There are no detailed ethical rules which might detract from free response to the kenotic ideal, and there is a minimum of divine action in the world – no more than the giving of visions of the good to encourage moral behaviour. Such is the minimum structure suggested by Ellis for the overall axiomatic purpose.

As he points out, a variety of religious traditions are consonant with the experienced nature of our many-levelled world and the assumption of an underlying purpose. They would all centre on belief in a transcendent loving moral reality (God), espouse the worth and value of each person, and advocate loving sacrificial behaviour. However, it is Christianity that emphasises *kenosis* (whatever the failures of its adherents), and he quotes William Temple's vivid phrase, 'power in complete subordination to love', as a summing up of the Creator's work.[359]

We ought to bear in mind Ellis's main aim. He deliberately limits himself to the task of linking the moral and physical realms, using a carefully explained scientific approach to effect a convincing integration. He keeps his model as simple as possible by idealizing it in accordance with well established scientific practice,[360] concentrating simply on the integration of kenotic morality into the scientific world-view.

He has thus inferred a theistic cosmology from part of the range of human experience and Christian tradition but, for the sake of simplicity, omits vital aspects such as our nature as dependent beings facing pain and death, who search for ultimate meaning and long for true beauty. He remarks that the much more difficult task of integrating aesthetic experience into the model has yet to be attempted. Furthermore, despite his strong endorsement of Temple's *Readings in St John's Gospel*,[361] he has deliberately left aside the question of divine revelation; that is to say, he stays within the bounds of natural theology.

Is there, perhaps, a simpler yet more profound axiom, or set of axioms, that not only calls for a moral way of life but also takes account of the strange, humanly unimagined juxtaposition of divine power and humility – power in complete subordination to love – and the full range of our ultimate questions?

TOWARD A TRINITARIAN COSMOLOGY [362]

It seems reasonable to suppose that if this magnificent creation has been designed to bring forth *kenosis*, that very quality must lie at the heart of the divine nature. Can we then begin to create a metaphysical scheme that will embrace the entire complexity hierarchy (including the level of the divine) in all its *being* and *becoming*? Following Ellis, we could look for an axiom, or set of axioms, that would allow our scientific world-picture to be linked to the Christian understanding of God as Trinity. Thus we begin with the idea not simply of a *kenosis bearing universe* but with that fundamental tenet of Christian belief, the *kenosis of God.*

In his superb exploration of the nature of creativity and its costliness – shown above all in the ongoing creation of the world – W H Vanstone devotes a chapter to the 'kenosis of God',[363] a phrase that contains something of the limitlessness, the vulnerability and the precariousness of authentic love. The universe – the totality of being for which God gives himself in love – is costly to the creator. Taking the graciousness of God as axiomatic, he argues that such graciousness does not hold back any reserves of power or wisdom or love. All is poured out into the creating and sustaining of the world and the bearing of all consequences. This suggests, then, the axiom:

The Lord God creates with utmost love

which is consonant with Temple's notion of power in complete subordination to love and gives room to the traditional Christian doctrine of God as both source and sustainer of the created order.

But is such a single axiomatic statement sufficient? Is it too inclusive, covering all manner of cosmologies? Before attempting to answer this, we consider briefly the notion of restricting the range of applicability of a theory, a good example of which occurred in the development of particle physics in the early 1970s, based on principles of symmetry. Steven Weinberg explains it thus:[364] In this kind of fundamental physics we do *not* want a theory that is as flexible as possible, capable of describing all imaginable kinds of force among the particles of nature. Rather, we hope for a theory that rigidly will allow us to describe only those forces that actually, as it happens, exist. Now, symmetry principles can give rise to a variety of theories, many of which are complicated. When, however, the very precise restriction is introduced that *whatever infinities arise in the calculations must all cancel*, this is found to impose a high degree of simplicity on the equations. Together with careful use of symmetry, this restriction went a long way to giving a unique shape to the 'standard model' of elementary particles. Physicists instinctively felt that thereby they had acquired a tighter grasp on that part of physical reality.

As in the case of particle physics, we do not want a highly flexible metaphysical 'theory of everything', capable, perhaps, of forming some sort of fit with most of the major religions of the world. Such a theory would be too general to have much meaning. Therefore we look for a restriction or modifying axiom that would make for a much closer fit with the insights of a particular religious tradition – Christianity in this case – a statement that follows on from the above axiom. For example, to allow for more than mere neural promptings towards kenotic behaviour,

> *The Lord God reveals (the divine nature and purpose) with utmost love*

and, to include the idea of not only the *being* but also the *becoming* and ultimate *fulfilment* of the created order,

> *The Lord God perfects (the creation) with utmost love,*

in line with the affirmation, 'Behold, I am making all things new'.[365]

If we allow our three-fold axiomatic declaration to conjure up a vision of tender divine invitation and the awe and joy of creaturely response to a radiant beauty crowned with thorns and crucified, this would bring us toward the heart and meaning of Christianity's Trinitarian doctrine of God. Such a vision can, reflexively, illuminate and add profound depth to the three statements.

Of course, ultimate reality may not succumb so easily to human summarizing. Nevertheless, here is a framework within which to place not only Ellis's postulate of a kenotically based Cosmology but also Christian belief about God as kenotically involved in the world, responding to the needs of its creatures, suffering in it and with it, and bringing it all to fulfilment. Paul the Apostle expresses the expectation of such a destiny when he writes that 'the creation itself will be set free from its bondage to decay and will obtain the freedom of the glory of the children of God'.[366]

What is surely needed in the light of today's knowledge is the reconsideration of the Church's doctrines of redemption and creation, and the relation between them.[367] If the Lord God creates, sustains, reveals, redeems and perfects in the totality of the cosmic drama – exercising power that is always in complete subordination to utmost love – this gives a richer meaning to the word *Gospel*. Perhaps it can be seen not simply as the *Gospel of Salvation* but, more broadly, as the *Gospel of Creation* (with *Salvation*, of course, as the crucial core).[368]

Constructing such a Trinitarian metaphysics is by no means a *foundationalist* project. Any search for a totally secure basis on which to construct knowledge all too easily encourages the idea that theology is based on a set of unchallengeable propositions. John Puddefoot argues that this is a misleading idea

> because it treats axioms, whether in mathematics or theology, as givens, as if axioms themselves have no history. ... This locates the roots of systems in quite the wrong place: an axiom is not the foundation of a system, but the product of generations of mathematical enquiry as it has eventually been formalized and *axiomatized*.[369]

Thus our set of three axiomatic statements for the construction of a trinitarian cosmology arise *a posteriori* from the Church's well-winnowed yet provisional and multi-faceted doctrine of God.[370] They are concise abstractions based on a rich variety of religious experience and worship. They provide a conceptual framework in which to look for further connections between our understandings of God and the universe, concepts such as *relationality* and *particularity* that would illuminate the meaning and purpose of the universe and its beings. But we turn first to two particular Christian doctrines that have been treated briefly in recent science-theology discussions.

THE BEGINNING OF HUMANITY AND THE FULFILMENT OF THE UNIVERSE

Several years ago the Jesuit theologian Christopher Mooney wrote a notable review of the possibilities of the science-theology interaction, in which he expressed the Church's need for a theological meta-narrative.

> The universe that science studies is not a mere sequence but a story, a struggle upward through matter, life, thought, history, and culture. Only a narrative can really capture what is going on. And it is precisely this need of humans for meaningful narrative that allows theology to complement the causality of science.[371]

Conversely, science's understanding of the human person as coming into existence through the processes of biological evolution has created the need to reformulate certain traditional Christian doctrines. Mooney continues:

> Christology, original sin, redemption, the theology of death, and the material character of the afterlife are the most obvious areas raising new questions which theologians must somehow confront.[372]

Here we limit ourselves to a discussion of original sin and the idea of eternal life, emphasising the latter since, at times, there is a reluctance in science-theology circles to address it.

Science-theology discussion, it should be noted, tends to be of the 'bottom-up' style that is characteristic of science – the way of 'universe-assisted logic'.[373]

> We look to evidence for what we are asked to believe. Bottom-up thinkers proceed from the basement of phenomena to the superstructure of theory. Top-down thinkers somehow seem to start at the tenth floor and to know from the start what are the general principles that should control the answers to the enquiry. Many theologians appear to the scientists to be of the top-down variety. As they discourse on the immanent Trinity, one wonders how they know what they are claiming. Bottom-up thinkers prefer to stick with the economic Trinity.[374]

Science gives us an account of the emergence of *homo sapiens* as the present culmination of an extraordinarily productive but often harsh process of evolutionary development. This raises acutely the question of how we are to understand the Christian doctrine of the Fall. For all who are involved in the constructive engagement of science and theology, the

Genesis narrative of Adam and Eve is a dramatic portrayal of Everyman (the root meaning of 'Adam') in his journey from innocence to responsibility and sin. It is not tenable as an account of a particular human being, but describes mythically[375] the profound alienation of humanity – from God, from other people, from our true selves, and also, writes Ian Barbour, from nonhuman nature, by denying its intrinsic value and violating our interdependence. Sin in all its forms is a violation of relatedness.[376]

The notion of original sin does not clash with science. In fact a link is sometimes suggested between that doctrine and the self-centred struggle for survival assumed in the theory of biological evolution.[377] Polkinghorne relates the Fall to the dawning of consciousness. It is conceivable that this faculty developed gradually in the line of higher animals and the hominids, and was followed by the growing self awareness of *homo sapiens*, bringing with it a spiritual awareness of the presence of God. Perhaps a struggle developed within the hominid psyche between response to God and the assertion of independence – a succumbing to the temptation to assert human autonomy over creaturely dependence. Such a Fall would have meant the turning of death as a simple matter of fact into a mortality to be feared. Awareness of humanity's transience combined with the sense of alienation from God would have brought anxiety and bitterness.[378]

What *is* problematical in traditional Christian thought about the Fall is the idea that the entire realm of nature was infected and distorted by the sin of Adam and Eve, for biology and paleontology indicate that life on earth has always involved a harsh struggle for survival, before the arrival of *homo sapiens* as well as after. Indeed, physical death is perhaps an essential part of God's evolutionary process of creating new forms of life,[379] to say nothing of its significance as the *eschaton*,[380] giving to existence a responsibility and seriousness that it could scarcely have otherwise.[381] Commenting on the vexed question of the pain and suffering of a world created by a compassionate God, Holmes Rolston makes the point that 'natural history is cruciform' – that kenosis is exhibited at all levels of life, but that the axiomatic idea of infinite divine love implies that God suffers too, in and with the creation.[382]

As for the long term future of the material world, the evidence from the scientific side indicates that *homo sapiens* and all other species will eventually become extinct and that, much further on, the material of the universe will disperse inexorably into cold oblivion. It was this prospect which prompted Steven Weinberg (physicist and Nobel laureate) to write the following lines as he looked down at the earth during an air flight:

> It is very hard to realise that all this is just a tiny part of an overwhelmingly hostile universe. It is even harder to realize that this present universe has evolved from an unspeakably unfamiliar early condition, and faces a future extinction of endless cold or intolerable heat. The more the universe seems comprehensible, the more it also seems pointless.[383]

Our scientist-theologians offer other views about the ultimate future of the creation. Peacocke is inclined to keep eschatological speculation to a minimum. He stresses the utter faithfulness of God and the Christian hope of remaining eternally 'in God'. The 'divine end of human becoming' is encapsulated for him in the phrase of Irenaeus of Lyon: 'to make us what even he (Jesus the Christ) is'.[384] Barbour seems to incline more to a process view of eschatology, summarizing thus the ultimate value of the created order:

> Every entity is valuable for its ongoing contribution to the life of God. The values achieved in this world are preserved in God's eternal life, and this is part of their enduring significance and permanence beyond the flux of time. In addition, some entities, such as human beings, have a (further) future value, if as conscious individuals we survive death.[385]

The full hope of process eschatology seems unclear but there is at least the expectation that our lives are meaningful because they are preserved everlastingly in God's experience.[386]

Among the scientist-theologians it is Polkinghorne who has written in the greatest depth about the eschatological scene,[387] which is a major concern of his for, as he says, a credible eschatology is essential for the coherence of Christian belief and is indispensable to theology.[388] Whereas Barbour presents the idea of each person or entity being preserved in the memory of God, Polkinghorne takes this much further in terms of the biblical hope of resurrection, which means not the *survival* of an immortal soul but *resurrection* of the human person as a new embodied being.

> The Christian hope is of death and *resurrection.* My understanding of the soul is that it is the almost infinitely complex, dynamic, information-bearing pattern, carried at any instant by the matter of my animated body and continuously developing throughout all the constituent changes of my bodily make-up during the course of my earthly life. That psychosomatic unity is dissolved at death by the decay of my body, but I believe it is a perfectly coherent hope that the pattern that is me will be remembered by God and its instantiation will be recreated by him when he reconstitutes me in a new environment of his choosing. That will be his eschatological act of resurrection. Thus, death is a real end, but not the final end, for only God himself is ultimate.[389]

Now the love of God is surely not directed simply to human beings, or even to the wider group of living creatures, for 'God saw all that he had made, and it was very good/beautiful'.[390] Therefore the physical matter of this world, imagines Polkinghorne, will itself be transformed into a resurrection world and there provide the bodily material for resurrected life, for God will no more abandon the universe than he will abandon us. This idea, he mentions, is akin to the Orthodox Church's understanding that the ultimate destiny of the whole creation is *theosis* (deification).[391]

Referring to the empty tomb as a hint of that total transformation, he continues:

> Hence the importance of the empty tomb, with its message that the Lord's risen and glorified body is the transmutation of his dead body. The resurrection of Jesus is the beginning within history of a process whose fulfilment lies beyond history, in which the destiny of humanity and the destiny of the universe are together to find their fulfilment in a liberation from decay and futility.[392]

Clearly there is much in the eschatological picture that is speculative, but it is reasonable speculation, in keeping with the axiom that *the Lord God perfects the creation with utmost love*. It also leads to the question: Given the costliness and pain involved in this vale of tears, why did God create it in the first place rather than making the 'new creation' directly? Polkinghorne sees it thus:

> (T)he new creation is not a second attempt by God at what he had first tried to do in the old creation. It is a different kind of divine action altogether, and the difference may be summarized by saying that the first creation was *ex nihilo* while the new creation will be *ex vetere*. In other words, the old creation is God's bringing into being a universe which is free to exist 'on its own'... (while) the new creation is the divine redemption of the old. .... The understanding that this creates in my mind is that the old creation has the character which is appropriate to an evolutionary universe, endowed with the ability ... to make itself. The new creation represents the transformation of that universe when it enters freely into a new and closer relationship with its Creator, so that it becomes a totally sacramental world, suffused with the divine presence.[393]

Keith Ward adds to this picture:

> It might even be that God's own nature, as love, is only fully realised by the creation of other conscious agents with whom God can share in fellowship, by giving, sharing and receiving a love that binds creator and creatures together in a community of spiritual being. If that is so, it is natural to hope that such a community might make it possible for every created member of it to share in knowledge of its final fulfilment in God. In other words, the love of God might require that the fulfilment of creation is not only experienced by the one consciousness of God, but shared in a communion of love that God brings to completion. In this way the existence of a resurrection world, however exactly it is envisaged, comes to seem a natural hope for a created cosmos. [394]

God reveals the existence of that resurrection world, he continues, in the life, death and resurrection of a particular human being in a carefully prepared historical context, thereby expressing the divine will to achieve the final fulfilment of every part of the creation. Altogether it is a picture that resonates with our three-fold axiom: God *creates*, *reveals*, and *perfects with utmost love*.

THE DOCTRINE OF CREATION IN THE LIGHT OF BEAUTY [395]

The belief that the creation is an expression of *utmost love* links readily to the idea that in the being of God the qualities of *truth*, *goodness* and *beauty* – the traditional triad of transcendental values – find their fullest and most perfect expression (which is a central theme of Richard Harries' *Art and the Beauty of God)*. Then *utmost love* would surely rejoice in the emergence of these values within the creation – within the life of human beings as made in the image of God.

So far we have been concerned with various questions of truth and reality, and the attempt to understand something of the universe and its Creator. We have noted the idea of the universe as a vast and costly enterprise – this 'cradle of human existence' that seems to be the bearer of meaning and value and gives intimations of the God of utmost love. Thus, we have dealt to some extent with the first two of these transcendentals but have hardly considered the third. Indeed, truth and goodness have always been at the forefront of Christian thought and concern, while beauty has often been ignored.

The philosopher-theologian Patrick Sherry writes:

> (T)he two terms, the 'Holy Spirit' and 'beauty' ... both represent underdeveloped areas of theology taken in themselves, let alone in terms of the connection between them ... The lack of a theology of beauty, both of beauty in general and of divine beauty in particular, follows in part from fear and suspicion of the question, expressed in pejorative terms like 'aestheticism' and 'elitism'. At best, beauty has often been treated as a Cinderella, compared with the attention paid by theologians to her two sisters, truth and goodness, an attention manifested in theology's predominant concern with doctrine and ethics, and resulting in the intellectualization of religion in recent centuries.[396]

In this section we note first the suggestion that the neglect of beauty is one aspect of a general malaise in the Western world, for it amounts to no less than the fragmention of modern culture. This is followed by a brief account of two theological responses in terms of the role of the Holy Spirit in the creation.

What struck Bishop Lesslie Newbigin most when he retired to England in 1974, having lived in India for over thirty years, was the disappearance of hope.[397] Writing in the 1980s, he compared this situation to the decay and disintegration of classical culture when it ceased to provide a meaningful framework for human life, leaving a wide space for a new metaphysics – a 'post-critical philosophy' that was articulated above all in St Augustine's early 5th century work *The City of God.* Newbigin endorsed Michael Polanyi's appeal for a new 'post-critical philosophy' as we stand at what feels like the end of a period of extraordinary brilliance.[398] Indeed, as a semi-retired elder statesman of the ecumenical movement, he spent the last two decades of his life (1978-1998) encouraging the Church to engage, like Augustine, in the enterprise of re-thinking the implications of the gospel for our public life and to offer its world-view to modern society as a convincing explanation of human existence. In effect he called for a new doctrine of creation, based on careful reflection of what it means to *know.*[399]

The impoverishment of modern culture is addressed by British theologian Colin Gunton in the first half of his book, *The One, the Three and the Many: God, Creation and the Culture of Modernity*. He quotes Alain Finkielkraut's summing up of present Western culture:

> We live in an age of *feelings.* Today there is no more truth or falsehood, no stereotype or innovation, no beauty or ugliness, but only an infinite array of pleasures, all different and all equal.[400]

The present malaise, comments Gunton, is a reflection of the constant pressure towards social homogeneity – towards the denial of the uniqueness of things and persons in all their material particularity – and he traces it to Plato's undue elevation of the rational element of human *being* at the expense of the aesthetic and the material. The implication is that we truly *are* when we think, but not when we love or make music. The universality of reason is set against the diverse particularities of the material, the *one* against the *many*. Since the arts engage with the material world in all its brute particularity and intractability, they are disqualified from being the bearers of truth and Plato could find no place for them in his ideal state. With the continued embodiment of such an outlook in the thought and practice of the Western world, including much of its theology, that world has never recovered from what Gunton (like others) refers to as the 'fragmentation of culture'. It has been in thrall continually to a doctrine that the *one*, but not the *many*, is of transcendental status.[401]

In the second half of his book he sets forth a detailed theological response. He explains that his thinking about God was deepened by the writings of the early 19th century poet and thinker Samuel Taylor Coleridge who, on a number of occasions, described the Trinity as 'the idea of ideas' – a notion that is extremely fertile in the generating of transcendentals (marks of all being) and related ideas, such as the contrasting but complementary concepts of *perichoresis*[402] and *particularity*. These refer to unity and distinctiveness, respectively. Like the concept of *kenosis*, they form useful springboards for further thought at the different levels of complexity and meaning.[403]

He deals first with the long-standing question of the balance between unity and diversity – in human society and in the human cultural enterprise. Can the three classical realms of culture – truth, goodness and beauty, or in the language of praxis, science, ethics and art – be related in such a way that the distinctive character and importance of each may be guaranteed, without elevating it above the others? Together, the three realms contribute to a rich and diverse social culture. In theological terms what is needed, suggests Gunton, is a renewal of the doctrine of creation based on a doctrine of God which in some way writes *plurality* into the being of things. So he looks for a way to conceive of the qualities of *perichoresis* and *particularity*, the *one* and the *many*, as equally important aspects of gracious divine creation.[404]

After discussing the idea of reality as a dynamism of relatedness – perichoretic on all its levels – Gunton turns to the notion of *particularity*. He considers that a trinitarian God whose creation reflects something of the rich plurality-in-relation of his being can surely enable us better to conceive something of the unity-in-variety of human culture.[405] Here he introduces an idea of Basil of Caesarea (c330-379), namely, that the distinctive function of the Holy Spirit is to bring to completion that for which each person and thing is created. But he goes a step further. If the orientation of the Spirit is to the *particular*, maintaining and affirming the particularity of each individual, then, suggests Gunton, perhaps that is the Spirit's role even within the Godhead – the particularity of created beings is established by the particularity at the heart of the being of God.[406]

Of the early Fathers of the Church it is Irenaeus of Lyon (c130-200) who does most to celebrate the goodness of the created world and, as Gunton puts it, few later theologians have achieved so adequate an integration of time and eternity, the one and the many. Patrick Sherry too draws upon the thought of Irenaeus, especially the concept of the Holy Spirit as the One who communicates beauty and inspires its creation. Irenaeus identified the Word of God with the Son and the Wisdom of God with the Spirit. God made all things by the Word, said Irenaeus, and adorned them by Wisdom.[407]

Introducing the subject of beauty in creation, Sherry writes:

> The full development of this idea involves the claims that the Spirit of God communicates God's beauty to the world, both through Creation, in the case of natural beauty, and through inspiration, in the case of artistic beauty; that earthly beauty is thus a reflection of divine glory, and a sign of the way in which the Spirit is perfecting creation; and that beauty has an eschatological significance, in that it is an anticipation of the restored and transfigured world which will be the fullness of God's kingdom.[408]

He also quotes extensively from the writings of the 18th century New England theologian, Jonathan Edwards, for whom the work of the Spirit is to quicken, enliven and beautify all things.[409]

So, bearing in mind John Taylor's broad picture of the Holy Spirit as *The Go-Between God* who is ever at work on the inside of creation, in the processes of nature, in history and in human living, we may speak of the Spirit as the One who, within that picture, affirms the particularity of each created being, seeking to bring it to the fullness of perfection and beauty. And borrowing from physicist Freeman Dyson's concept of a 'principle of maximum diversity' operating in the world, we may think of the work of the Spirit in terms of a 'principle of maximum beauty', making for the expression of beauty in its wide variety of forms.

BEAUTY ON EVERY SCALE

The nature of beauty ranges widely over a broad spectrum of associations – from the elegance of a scientific theory, to harmony in nature or art, to the lively inner beauty of a human being, and then to its ultimate form as the radiant 'terrible beauty'[410] of the glory of God. It seems that the higher the level of complexity and meaning, the more intimately and perichoretically is beauty linked to both truth and goodness. And because it is thus linked, writes Richard Harries, it can sometimes seem wracking and awesome. Thus a single English word is used for the entire range of this transcendental quality that is associated with the power to move and attract profoundly.[411]

But there is a different way in which beauty can manifest a wide spread, as pointed out by Nobel laureate and astrophysicist Subrahmanyan Chandrasekhar who often pondered the links between truth, beauty and scientific creativity. He believed that the aesthetic aspect forms a fundamental link between the arts and the sciences, and raised the question of how one may evaluate scientific theories as works of art. In science, as in any work of art, one looks for the proper conformity of the parts to one another and to the whole.[412] He goes on to quote the remark of the sculptor, Henry Moore, that the greatest sculptures, such as those of Michelangelo, should be viewed from all distances since new aspects of beauty will be revealed on every scale – and then explains that this is how Einstein's Theory of General Relativity shows supreme beauty, for it is strangely apt at any level in which one may explore its consequences. If this idea applies to works of art and scientific theories, it must surely be of profound significance for the conveying of religious belief.

Christian apologists, in particular, are always in need of a theological 'theory of everything', the full story of creation. Scripture urges: 'Always be ready to make your defence to anyone who demands from you an account of the hope that is in you'.[413] This is a challenge which can be met at many levels, from simple personal narrative to a philosophically rigorous dissertation. But Chandrasekhar's pointer to the beauty that can be revealed *on every scale*, suggests an even greater challenge to the Christian Church, or to any other community that seeks to commend its beliefs. If the Church is to be an icon of the truth, goodness and beauty that are the incomparable marks of the divine nature, it will surely be most persuasive when on every scale and in every aspect of its life and thought – from personal friendship to liturgy, from private prayers to doctrinal formulations, from artistic creation to politics – it shows forth something of the poignant beauty of existence in this costly creation.

Then if, as Harries puts it, God is beauty as much as he is goodness and truth – if in the being of God our three transcendental values are perichoretically linked – the aim to show forth beauty on every scale must go hand in hand with the quest to know and understand at all levels. And such a quest would surely form the basis of any effective search for what Dietrich Bonhoeffer called the 'non-religious interpretation of Biblical concepts'. This phrase constitutes his tantalisingly brief summary of an open ended theological programme to grasp and to declare the contents of the Christian gospel in such a way that this would lead to a new synthesis, with fresh metaphysics, in the language and ideas of our modern world.[414]

6 FUTURE AGENDA

One way of summarizing the broad range of present (and perhaps future) concerns in the field of Science & Religion is to place them under headings and subheadings such as those tabulated on the next page. Some of these have long been the subject of study and debate, perhaps with a history that needs to be understood, while others are listed simply as potentially fruitful topics.

The main thrust of the science-theology discourse over the past two or three decades has been the engaging of theological thought with successively higher levels of the complexity hierarchy – physics and cosmology, then biology and its complex systems, and more recently, studies of consciousness and other aspects of human being and becoming. Such searching for conceptual understanding will surely continue to develop further, especially in the area of artificial intelligence. From the reductionist side of the debate in this area comes the bold expectation that 'crossing the threshold of self-understanding is well within the reach of human intellect and ingenuity .. a full working understanding of the essential details of conscious physical processes may approach surprisingly soon'.[415] A more complete understanding would certainly require insights from higher levels of the hierarchy. On other scientific fronts, there is no end in sight in the search to understand the world of quantum

Table 4. Topics in the Science & Religion Field

Our Multi-Levelled Universe
- development of complexity in nature
- the realm of quantum phenomena
- mind, consciousness and free will
- genes and the evolution of culture and ethics

Natural Theology
- epistemological questions
- scientific theories and 'theory of everything'
- divine action
- theological grand narrative

Science, Theology and Ethics
- biotechnology and medical science
- technology and environmental science
- ecological theology (ecotheology)
- consequent ethical questions

Interfaith Exploration
- metaphysics: truth, goodness and beauty
- the nature and role of scripture
- the nature and destiny of the human being

Christianity and the Scientific World-Picture
- rethinking central Christian doctrines
- 'non-religious interpretation of biblical concepts'
- theology of beauty

phenomena and its constituents, the nature and process of the cosmos in its earliest moments, and the emergence of the first organic cells to form the basis of terrestrial life. The search may conceivably lead to yet undiscovered holistic principles which facilitate the emergence of life and its growing complexity.

The broadening of the discourse into the realm of ethics and political decision-making continues. In the areas of rapid technological advance, especially biotechnology and the vast realm of information technology, new ways of knowing and living will certainly call for wise choices. Genetic engineering and cloning are already topics of intense debate. It seems that these developments imply an ever increasing capacity to influence the world of animal and plant life and even, perhaps, the entire global environment. As it is, there is much confusion about the relation between humankind and the remainder of the natural world, and voices have been raised about the urgent need for the development of an ecological theology.[416]

In the area of human relations there is a potentially creative role for science-theology discussion, as a neutral meeting place for interreligious exploration. There seems to be a significant openness to such dialogue amongst theologians and in some of the mainstream thinking of the Christian Church in the West.[417] This is not to deny the differences of approach between the conservative and liberal wings of the Church, or within its evangelical ranks, but the rethinking of theology's story of creation, without diminishment of its core christological doctrine, may allow the differences to be softened – even to be affirmed – and make for a far greater inclusivism.[418] From the Christian side of the exploration, a view of the Holy Spirit as the one who maintains particularity and who inspires beauty (in art as well as nature) would surely be especially relevant to the task of creating interfaith understanding.

The experience gained in the science-theology discourse, in both its historical and epistemological aspects, could help the Church in the West to think and act in an appropriately affirmative manner towards the indigenous peoples of the world and their cultures, taking account of the sharp contrast between scientific knowledge and what the postmodern philosopher Jean-Francois Lyotard calls narrative knowledge – popular stories, myths, legends and tales.[419] The situation calls for openness to the potential contribution of such indigenous wisdom and ways of life.[420] A new theology of nature would find here some striking examples of the celebration of life in a world that is viewed holistically and with a profound sense of dependence and belonging.

The Science & Religion movement is a burgeoning enterprise, taking the form of a wide variety of conferences, articles, books, occasional television documentaries, and hundreds of tertiary-level courses in North America, Europe and further afield. It is accompanied by a growing interest in the debate which will no doubt help to eliminate the caricature of the science-religion relationship (often nurtured by the media, as one of unremitting conflict) and allow a more sophisticated public discussion to develop.[421] Altogether it will continue to be a vital part of humanity's search to know and understand.

ENDNOTES

1 Edward Grant in Lindberg & Numbers 1986:52

2 *Metaphysics* deals with a very general picture of what the world is like, what things there are in it and what we can know about it. It engages in the attempt to know reality, not mere appearance. *Logic* is the theory of reasoning, proof, thinking, or inference. *Epistemology* is the study of the nature of knowledge; how and to what extent we possess different kinds of knowledge.

3 *World-picture* refers to the mental model of the structure and processes of the universe, whereas *world-view* implies an evaluation of the world – whether, for example, it possesses meaning and purpose.

4 Roughly the period from the mid-15th to the early 17th century.

5 Polkinghorne 1988:xii

6 ibid, pp 9-10

7 E Zilsel, 'The Genesis of the Concept of Physical Law', in *Philosophical Review*, vol 51 (1942), pp 245-279, discussed in Cohen 1994:453
8 A purely reason-based belief in an absent Creator God. See the section on 'Early Natural History and the Emergence of Deism'.
9 Goodman 1973:319
10 Prevalent first in Germany and later in Victorian England through the development of techniques of biblical criticism.
11 Discussed further in the section on 'Epistemic Aspects'
12 C Mooney 1991, 'Theology and Science: a new commitment to dialogue' in *Theological Studies* (Georgetown U, Washington DC), vol 52, p 310
13 Apart from such affirmation of epistemic variety, postmodern thinkers are inclined to dismiss the notion of absolute truth or reality and to reject any claim to absolute foundational knowledge.
14 Polkinghorne 1996:14
15 ibid, p 1. Polkinghorne calls it the great integrating discipline. It should be noted, however, that within the science-theology debate there is some sympathy for postmodernism's rejection of all forms of grand narrative with their threat of epistemic imperialism, and for the idea that the search for truth is better served by holding together a variety of perspectives.
16 The word 'planet' is derived from the Greek word *planetes* (wanderer), a name prompted by the seemingly erratic paths of these five bodies.
17 Butterfield 1957:6
18 ibid, p 7. However, it was the realm of biology that most interested Aristotle; he was more concerned with the changes that occur in *living* things than the changes represented by mere mechanical motion or process. Indeed, there is a tendency now to think that he was an excellent naturalist.
19 Rowse 1972:227
20 Astrology is the art of determining the influence of the stars and planets on human affairs. Alchemy, the medieval forerunner of chemistry, was especially concerned with the attempt to change the baser metals into gold or silver by evoking secret magical powers of nature.
21 Butterfield 1957:35
22 Cohen 1994, *The Scientific Revolution: A Historiographical Inquiry*

23 ibid, p 69
24 Hooykaas 1972:75
25 Cohen 1994:86
26 ibid, p 75
27 Butterfield 1957:vii
28 Steven Shapin begins a book with the provocative assertion: 'There was no such thing as the Scientific Revolution, and this is a book about it'.
29 Cohen 1994:5
30 Student of Chinese society and its history.
31 Cohen 1994:16, also pages 378-488 (on 'The Non-emergence of Early Modern Science outside Western Europe'). Needham himself was a Cambridge University biochemist who became a renowned scholar in the history of Chinese science and culture.
32 ibid, pp 438, 431-437
33 Cohen 1994:438
34 ibid, p 382
35 ibid, p 431
36 ibid, p 385-6
37 ibid, p 386, and Brooke 1991:43
38 ibid, p 385
39 ibid, pp 398-401. Here Cohen draws upon the essay 'The Causes of the Decline of Scientific Work in Islam', an appendix to Sayili's *The Observatory in Islam and its Place in the General History of the Observatory* (Ankara, 1960).
40 Drawn mainly from Cohen 1994:378-488.
41 von Grunebaum in Cohen 1994:391.
42 Cohen 1994:417
43 ibid, p 407
44 ibid, p 408
45 from Nasr's *Man and Nature: The Spiritual Crisis of Modern Man* (1968), quoted in Cohen 1994:485-486. See also Cohen's footnote 81 on p 590.
46 M Negus in Southgate 1999:315-316
47 Cohen 1994:321-367
48 PhD thesis of Robert Merton, published as *Science, Technology & Society in 17th Century England* (1938). See Hooykaas 1972:98-149 and Cohen 1994:200-203.

49 Lindberg & Numbers 1986:4-5 and 192-217 ('Puritanism, Separatism, and Science' by Charles Webster). See also Cohen 1994:314-321 and Brooke 1991:109-116
50 Cohen 1994:254-9
51 Hooykaas 1972:75-76
52 Cohen 1994:486 (quoting from Needham et al, *Science & Civilization in China*, vol 2, p 431).
53 Bowker 1995:166
54 Newbigin 1986:70-72. A view derived in part from his many years of conversing with Hindu thinkers while a bishop in the Church of South India.
55 Lynn White in Cohen 1994:478.
56 Russell 1985:37
57 Explanations in terms of ultimate purpose.
58 Butterfield 1957:32
59 Plato taught that the physical entities of the world can best be understood as concrete expressions of fundamental Forms or Ideas which possess a quality of being and a degree of reality superior to that of the physical world (Tarnas 1991:6).
60 C Russell in Goodman & Russell 1991:52
61 Hooykaas 1974, Open U Course, Unit 2, p 63
62 For example, Barbour 1966:33
63 C Russell in Goodman & Russell 1991:60
64 Gingerich O, 'The Galileo Affair' in *Scientific American*, vol 247 (August 1982), pp 118-127.
65 Goodman & Russell 1991:65-69. Until the growth of historical-critical approaches to the Bible in the 18th and 19th centuries, Moses was generally regarded as the author of the Pentateuch, the first five books of the Bible.
66 Arthur Koestler in Wertheim 1995:70
67 Goodman 1973:8
68 Wertheim 1995:71, 74-75
69 Russell 1985:47-48
70 EA Burtt in Cohen 1994:96
71 Goodman 1973:32
72 S Drake in Goodman & Russell 1991:108
73 D Goodman in Goodman & Russell 1991:114. Note that three of the ten examining cardinals refused to sign the sentence.
74 Geoffrey Cantor in Poole 1995:113

75 That which was promulgated by the Council of Trent (1545-63).
76 D Goodman in Goodman & Russell 1991:115
77 Also, in the case of its response to Darwin's theory of evolution, the Catholic Church had come to terms with the theory well before Vatican II. But it was not until 1996 that the Pope announced publicly that it was time to recognise the theory as 'more than a hypothesis', for any knowledge based on truth is not incompatible with our knowledge of God.
78 Cohen 1994:323
79 EJ Dijksterhuis in Cohen 1994:70, 71
80 Butterfield 1957:79
81 Brooke 1991:22; also Bernal 1965:441-2
82 Bernal 1965:442-3, 447
83 Cohen 1994:196-7
84 The following outline is drawn from Wertheim 1995:154-156.
85 Characterized by philosophical subtleties and precise definitions of dogma.
86 Sorell 1987:6-9
87 EJ Dijksterhuis in Cohen 1994:70
88 Goodman 1974, Open U Course, Unit 4, pp 13, 14
89 ibid, p 16
90 ibid, p 27
91 ibid, p 31
92 ibid, p 31
93 Cohen 1994:498; also Lindberg & Numbers 1986:137-142 and (on the mechanistic philosophy) 1986:279-286.
94 Brooke 1974, Open U Course, Unit 5, pp 49
95 Butterfield 1957:127
96 ibid, pp 131, 132
97 Wertheim 1995:100
98 Hooykaas 1972:139
99 Reproduced in Goodman 1973.
100 Goodman 1974, Open U Course, Unit 4, pp 36, 37.
101 Barbour 1966:37
102 Russell 1985:112, 113
103 Margaret Jacob in Wertheim 1995:129
104 Russell 1983:52-61 and 1985:124

105 Chandrasekhar 1987:64. This distinguished astrophysicist checked all the mathematics and was deeply impressed by the skill and subtlety shown.
106 Westfall 1980, chapter 8; also White 1997, especially chapters 6 and 7
107 Wertheim 1995:115
108 Westfall 1980:22
109 White 1997:206
110 White 1997:120 and chapters 6 and 7. The author writes: The hermetic tradition – the body of alchemical knowledge – was believed to have originated in the mists of time and to have been given to humanity through supernatural agents.
111 Tarnas 1991:487
112 Brooke 1974, Open U Course, Unit 5, pp 84 and 90-91
113 EA Burtt in Russell 1973:132.
114 Brooke 1974, Open U Course, Unit 5, p 87
115 EA Burtt in Russell 1973:133
116 Brooke 1974, Open U Course, Unit 5, pp 87-88
117 It took place from the mid-15th to the early 17th century.
118 Cohen 1994:286-296
119 Butterfield 1957:187
120 ibid, p 190
121 ibid, p 182
122 Tarnas 1991:282
123 Newbigin 1984:7
124 Gunton 1985:115-119. He writes on p 119: Where we employ concepts, they used pictures.
125 Barbour 1966:40
126 Neill 1958:181-2
127 Brooke 1991:146, 150-151
128 ibid, pp 7-8
129 Lindberg & Numbers 1986:12, 264
130 Goodman 1974, Open U Course, Unit 7, pp 52-54
131 Vidler 1961:13, 237
132 Gonzalez 1985:190, 193
133 Goodman 1974, Open U Course, Unit 7, p 42. Goodman writes: One writer of that period described a *deist* as the sort of man who was not weak enough to be a Christian and not strong enough to be an atheist.

134 Brooke 1991:152-3

135 Goodman 1974, Open U Course, Unit 7, pp 40-41

136 A *species* is a group of animals or plants capable of interbreeding. Since they cannot interbreed, dogs and cats are regarded as belonging to different species. Similar *species* together constitute a *genus*. Likewise similar *genera* are grouped into *families*, similar *families* into *orders*, similar *orders* into *classes*, similar *classes* into *phyla* and similar *phyla* into *kingdoms*. Our genus/species is homo sapiens, in the family of hominids, in the order of primates, in the class of mammals, in the kingdom of animals.

137 Blackmore & Page 1989:18, 20

138 The idea that nature comprises a series of organisms of all gradations, ranging from unicellular structures to human beings. It stemmed from the idea that God would allow no niche in nature to remain unfilled.

139 S Lindroth in Goodman & Russell 1991:323

140 Goodman 1974, Open U Course, Unit 7:55,59

141 See the section on Newton who thought of gravity acting as the 'puppet strings of God'.

142 Brooke 1991:234, 236-237

143 Hybrids, such as the mule, do not count since they are sterile.

144 Goodman, 1974, Open U Course AMST 283, Unit 7, p 59

145 J Wilkie in Russell (ed) 1973:246-247

146 Jansenism was a Calvinist-oriented political and intellectual movement within the Catholic Church in France (Gonzalez 1985:169).

147 Goodman 1974, Open U Course, Unit 7, p 57

148 ibid, p 56

149 ibid, p 58

150 The atheistic view that the world comprises simply atoms in random motion. (J Wilkie in Russell (ed) 1973:264)

151 Goodman, 1974, Open U Course, Unit 7, p 60

152 J Wilkie in Russell (ed) 1973:263-264

153 Blackmore & Page 1989:48,49

154 The idea of *autopoiesis* in the natural world – self-making by a process of self-organization – lies at the heart of today's research in biological evolution.

155 An *invertebrate* creature is one which does not have a backbone.

156 J Wilkie in Russell (ed) 1973:268. He thought an increase in size implies an increase in complexity.
157 ibid, p 275
158 ibid, p 270
159 Blackmore & Page 1989:50
160 Bowler 1989:112, 113, 114
161 Gillispie 1959:99-100
162 Two earth scientists at Columbia University (New York), William Ryan and Walter Pitman, have written *Noah's Flood* (1998) in which they present geological and archaeological evidence for a prolonged cascade of water as the Mediterranean broke through a natural dam in the Bosporus Strait and plunged into what was then a freshwater lake and is now the Black Sea. They claim that it surged in for at least 300 days, inundating 60 000 square miles of land. They date it at about 5600 BC and suggest it as a story for oral transmission between generations. (*Scientific American*, June 1999).
163 Russell 1985:135
164 Gillispie 1951:75
165 The account given in the first five books of the Bible which were assumed to have been written by Moses.
166 Gillispie 1951:50
167 ibid, p 103
168 ibid, p 105
169 Hooykaas and Lawless 1974, Open U Course, Unit 11, p 70
170 Martin Rudwick in Russell 1973:208
171 Gillispie 1951:106
172 Russell 1985:14
173 Gillispie 1951:108
174 Russell 1985:139
175 Brooke 1991:248-249
176 ibid, p 253
177 ibid, p 250,276
178 Hooykaas and Lawless 1974, Open U Course, Unit 11, p 78
179 Browne 1995:140, 187
180 Brooke 1991:251
181 Brooke 1974, Open U Course, Unit 12, p 38
182 ibid, p 38
183 Greene 1982:13,14

184 The plate tectonics theory was neatly confirmed and elaborated in the 1950s and 60s from detailed mapping of the entire seabed and the patterns of magnetism in its rock layers.
185 Dupree in Lindberg & Numbers 1986:354
186 Browne 1995:189
187 ibid, p 186
188 ibid, p 317
189 Darwin's *Origin* (ed Beer) 1996:xiii
190 Browne 1995:240.
191 Barbour 1966:85 and Russell 1985:146
192 Darwin's *Origin* (ed Beer) 1996:xv-xvii
193 The full title of the book was *On the Origin of Species by means of Natural Selection, or the Preservation of Favoured Races in the Struggle for Life*. Although the commonly used shorter title seems to emphasise the idea of origin, Darwin made no claim to studying how life began; he was concerned simply with its evolutionary process, not its origin, and certainly not with any notion of first causes.
194 Hopper 1987:66
195 Howard 1982:55
196 Darwin's *Origin* (ed Beer) 1996: viii, ix
197 Desmond & Moore 1992:623
198 Vidler 1961:112,113
199 Desmond & Moore 1992:657
200 Contemporary press reports quoted in Desmond & Moore 1992:668
201 Russell 1985:146; Darwin (ed Beer) 1996:xx
202 Darwin's *Origin* (ed Beer) 1996:39, 67-68
203 ibid, p xxii. Note that Spencer himself showed more of a Lamarckian than a Darwinist leaning in his new popular philosophy of universal evolutionism, postulating an essentially progressive process in human society. See Bowler 1993:238.
204 Blackmore & Page 1989:78-79. See also the detailed discussion in Bowler 1989:208-214. The problem was not satisfactorily resolved until this erroneous understanding of heredity was corrected through the emergence of Mendelian genetics in the early 1900s.
205 The radioactive decay of certain heavy chemical elements and the slow solidification of the molten iron core (with the release of 'latent heat').

206 Moore 1979:136

207 The introduction to each edition of Darwin's *On the Origin of Species* ends with the sentence: 'I am convinced that Natural Selection has been the most important, but not the exclusive, means of modification'.

208 Blackmore & Page 1989:71

209 The remarkable perpetuation of the conflict idea is treated in Colin Russell's paper, *The Conflict Metaphor and its Social Origins* (Science & Christian Belief, vol 1, No 1, p 3). Also Moore (in 1979:19) writes that through constant repetition in historical and philosophical exposition of every kind, from pulpit, platform, and printed page, the idea of science and religion at 'war' has become an integral part of Western intellectual culture. Like other clever metaphors, this one shows few signs of dying out. (See endnote 421 on the contrasting situation in Russia).

210 Moore 1979:194-196.

211 This aspect of science has been described by the early 20th century physicist, Ernest Rutherford, as 'listening to nature's whispers'.

212 Bowler 1983:15, 155, 215.

213 Blackmore & Page 1989:106

214 ibid, pp 109-110 and Moore 1979:82.

215 Moore 1979:303

216 Richardson suggests that the revolution in historical method in the period 1760-1860 was as important in the development of human thought and life as was the revolution in scientific method in the period between Copernicus and Newton. Elements of higher criticism had already been started by Hobbes and Spinoza in the 17th century and this approach to Scripture was developed from the mid-18th century onwards, especially in German theology (See Open U Course, Unit 14, pp 117-119, 122-124). Furthermore, the disturbance to Victorian religious belief reached a climax with the publication of *Essays and Reviews* (1860) in which seven Anglican contributors published their exploratory ideas on questions that had arisen in the new theological climate of the day. The book provoked a declaration against the spread of liberal theological views, signed by nearly 11000 clergymen and 137000 of the laity. (Vidler 1961:123 ff, 128).

217 Russell 1985:149

218 Moore 1979:103

219 The present-day Oxford biologist, Richard Dawkins, makes the same point when he speaks of nature as 'pitilessly indifferent' in its outworkings of natural selection, a process without direction or purpose other than to propagate the genes of one generation into the next.

220 Moore 1979:203

221 F Gregory in Lindberg & Numbers 1986:374

222 Moore 1979:220

223 ibid, pp 248-249

224 Russell 1985:164-165 and Moore 1979:224

225 H Dupree in Lindberg & Numbers 1986:361, 362

226 Russell 1985:167-168.

227 Moore 1979:253

228 ibid, p 254

229 ibid, p 259

230 ibid, p 260

231 ibid, pp 263-264

232 ibid, pp 268-269

233 Tarnas 1991:284-5

234 Ronald Numbers gives a comprehensive account in *The Creationists: the Evolution of Scientific Creationism* (1992, U California Press) and *Darwinism comes to America* (1998, Harvard U Press). See also Edward Larson's *Trial and Error: The American Controversy over Creation and Evolution* (1989, Oxford U Press).

235 Berry 1988:86. Professor Berry is a geneticist at University College London and a former President of the British organization, *Christians in Science*. See also Julian Huxley's *Evolution: the Modern Synthesis* (London: Chatto & Windus, new ed, 1963)

236 T Dobzhansky in *American Biology Teacher*, vol 35 (1973), pp 125-129.

237 Berry 1988:87-88

238 See also Blackmore & Page 1989:144-151. These mechanisms include the genetic drift already mentioned. Then there is the observation of the similar development of a given species in long separated geographical regions – for example, the Tasmanian wolf and the Canadian timberwolf, which evolved from the same shrew-like ancestor, suggesting that there may exist a basic pattern that provides limits to the genetic variations. Again, there is speculation about a possible neo-Lamarckian mechanism of 'reverse tran-

scription' in which changes in body cells (through usage or environmental pressures) might produce changes in the genes.

239 First proposed by SJ Gould and N Eldridge in *Paleobiology*, vol 3 (1977), pp 115-151; see also Bowler 1983:336-338.

240 The term used to describe the spreading out of a single species into new forms and species suited to the available environmental niches.

241 The *Second Law* (or *Law of Entropy*) states that, within any *closed* physical system, all changes make for a net increase of entropy – the arrangement of its constituent elements becomes less ordered. A *local* increase of order is permitted (in any limited region of the system), but only at the expense of a more than compensating amount of disorder in the remaining parts. The Second Law is sometimes invoked naively as an argument against the evolution of life from simpler to more complex forms, in disregard of the fact that no living thing is a closed (isolated) system.

242 These are now known to range from the highest energy cosmic rays, through x-rays, ultraviolet light, visible light, infrared radiation, microwaves, to the longest radio waves.

243 Wertheim 1995:4

244 Davies (ed), *The New Physics* (1989)

245 As a rough model one can picture a quantum of light (a photon) as a short burst of waves (a wave packet) which exhibits either a particle-like or a wave-like nature, depending on the way it is observed.

246 In present quantum theories the different types of inter-particle force take the form of exchanges of these go-between particles.

247 Isolated neutrons decay after a few minutes, but within an atomic nucleus they remain stable.

248 A term introduced to interpret an elegant experiment on the Indeterminacy Principle, performed in 1998. Physicist-philosopher David Bohm spoke of 'unbroken wholeness' rather than 'non-locality'.

249 Intelligibility, not picturability, is what gives physicists confidence in the reality of fundamental particles.

250 Brian Greene's *The Elegant Universe* (1999) provides a clear description of particle physics and its development in this latest form, which is often called simply 'superstring theory'. He writes (on p 226) that this theory has been hailed as the most important

and exciting development in theoretical physics since the discovery of quantum mechanics in the 1920s.

251 An asymmetry due to absorption and scattering of light by dust particles in our galaxy.

252 The speed is determined from the shift in frequency of the emitted light (as its wavelength increases with the stretching of space) and the distance is given roughly by the brightness. As an illustration, consider the expanding surface of a steadily inflated balloon marked with equally spaced dots. From any dot the recessional speed of any other dot will be observed to be proportional to its distance away. It is assumed that, likewise, the expanding universe would appear the same from the vantage point of any galaxy.

253 Determined, for example, from the ratios of certain isotopes of lead in the earth's surface.

254 This would not violate the Principle of the Conservation of Energy if the creation field carried negative energy.

255 Even though he came to the conclusion that the universe is 'a put up job', Fred Hoyle is known for 'the explicitness with which he introduces attacks on Christian religious beliefs into his cosmological works' (see Ernan McMullin's paper in Peacocke 1981, p 34).

256 The Big Bang seemed like evidence in favour of the idea of divine creation and, much to the consternation of some theologians, especially the Jesuit cosmologist Abbe Lemaitre, Pope Pius XII acclaimed it as such.

257 Chown's *The Afterglow of Creation* (1993) is an absorbing account of this scientific drama.

258 Determined by analysing light from intergalactic gaseous matter.

259 Ten multiplied by itself thirty times.

260 Weinberg's *The First Three Minutes* (1978) is an impressively argued description of the unfolding early universe.

261 Since the mid-20th century the concept of nuclear fusion – the uniting of two lighter nuclei to form a heavier nucleus plus energy – has been applied not only to understanding the burning processes that form the life-cycle of a star but also in the making of hydrogen bombs and in the multibillion dollar attempts over the past 40 years to harness fusion energy for commercial use.

262 The temperature unit here is the Kelvin (1 K = 1 degree C) and the number of such units represents 'degrees C above Absolute Zero'.

263 A system is *linear* if its response to some perturbing force is proportional to the force; it becomes *nonlinear* if the force is increased to the point of non-proportionality.

264 Gleick 1987:3-4

265 The description of such systems is what is aimed at in 'chaos theory'. Where a precise mathematical description can be given, the system is regarded as 'deterministic' insofar as it actually matches the description – but its behaviour is often so complex as to be *de facto* unpredictable.

266 Kauffman, *At Home in the Universe* (1995). See also Capra's chapter on 'Self-Making' in *The Web of Life* (1997), especially pp 194-202 and 221, and Stewart's *Life's Other Secret* (1998)

267 Kauffman 1995:26-27

268 The randomness of mutational change is often invoked polemically against the Darwinian idea of evolution via natural selection. Randomness implies harmful mutations as well helpful ones, it is argued, so how can evolutionary *progress* be assured? Julian Huxley claimed that it is the helpful mutations that are slowly and inexorably selected and passed on through the processes of heredity – see Young 1986:141. However, it is not clear that such selection on its own makes for increasing complexity and consciousness – see Kauffman 1995:183-189.

269 See Prigogine & Stengers' *Order out of Chaos* (1984). As mentioned earlier, although the *total* entropy (or degree of disorder) of a closed system always increases there can exist local regions of diminishing entropy (or increasing order) within the system.

270 Capra 1997:184-185

271 However, the mathematical laws may not model the system perfectly.

272 John Polkinghorne offers the slogan 'epistemology models ontology', suggesting that unattainability of knowledge of the unfolding of nature's processes is an indication (not by any means a strong one) that these processes are genuinely open and indeterminate. This controversial but metaphysically fruitful idea is discussed later.

273 Peacocke 1993:241. This is the thesis of Richard Dawkins' *The Selfish Gene* (Oxford U Press, 1976)

274 Capra 1997:77. See pages 92 and 228-9 of his book for speculation about the origin of the earliest forms of life – emerging via a prebiotic phase of evolution involving processes of 'molecular self-organization'.

275 Gleick 1987:5

276 Elegantly described in Brian Greene's *The Elegant Universe* (1999).

277 Hawking 1988:156

278 With new advanced techniques, astronomers are beginning to detect planets around other stars in our galaxy. As the techniques improve, it seems likely that thousands more will be detected over the next twenty years.

279 Polkinghorne 1983:8

280 See the detailed account in Davies' *The Cosmic Blueprint* (1987).

281 Did God have any choice in the way he created the world? asked Einstein, speaking of 'God' in the sense of Spinoza's use of the word, signifying the (non-personal) underlying rationality of the world.

282 The name Anthropic Principle is associated with the Greek word *anthropos* for 'human being'. There is a highly controversial version – the Strong Anthropic Principle – which asserts that the laws of nature *must* be such as to allow the existence of observers in the universe at some stage, the underlying assumption being that the idea of an observerless universe is meaningless. Barrow & Tipler's *The Anthropic Cosmological Principle* (1986) is the most comprehensive work on this topic. See also Paul Davies' *The Accidental Universe* (1982) and *The Mind of God* (1993), pp 161-193.

283 Constants such as the various particle masses and the various force strengths, and conditions such as the initial expansion rate of space and the amount of matter and energy in the universe.

284 Stuart Kauffman gives a high role to the self-organizing ability of molecular and organic systems and sees the emergence of life-forms as almost inevitable, leading to the creation of conscious beings of some sort, perhaps many sorts, not necessarily human-like and not necessarily limited to planet Earth. In contrast, the idea that the entire multilevelled evolutionary process of nature, cosmic and biological, is a delicate balancing act designed superbly and expressly to produce human beings is comprehensively treated in Michael Denton's book *Nature's Destiny* (1998).

285 Davies 1993:16
286 Polkinghorne 1986:51
287 Monod 1972:110, 167
288 Polkinghorne 1996:46
289 Polkinghorne 1986:54
290 Peacocke 1979:95, 117
291 This is perhaps linked to the longstanding question whether mathematics is *invented* (as the free abstract creation of the human mind) or *discovered* (as eternally existing truths), or a combination of the two.
292 Wigner in *Communications in Pure and Applied Mathematics*, vol 13, (1960) p 1.
293 The complexity hierarchy is treated briefly in Peacocke 1993:22, 36-41 and explored in great depth in Murphy & Ellis' *On the Moral Nature of the Universe* (1996).
294 It has been suggested that this renaissance began with Ian Barbour's comprehensive survey, *Issues in Science and Religion* (1966). However, note the annual Gottingen Conversations that took place between physicists and theologians from 1948 to 1959 under the leadership of mathematician Gunter Howe, described in Nebelsick 1981:159-177. These formed the basis for ongoing discussions which were taken up by the World Council of Churches and culminated in their influential conference at the Massachusetts Institute of Technology in 1979 on 'Faith, Science, and the Future'.
295 Edward Grant in Lindberg & Numbers (1986), pp 52-53.
296 ibid, p 54. Some natural philosophers used Aristotelian principles to deny the divine creation of the world, or that God could create more than one world, or that God could move the world in a straight line, etc.
297 ibid, p 53
298 Mooney 1991:308 (ref: note 12). Also note that in the USA during the 1990s there has been little traffic between those involved in Science & Religion courses and conferences and those seeking a new Christian apologetic through, for example, the church-mission orientated Gospel & Our Culture Network.
299 ibid, p 312. Note, for example, that in the 1998 General Assembly of the World Council of Churches (in Harare, Zimbabwe) no explicit attention was given to Science & Religion concerns.

300 From Polkinghorne (*Science & Christian Belief*) to Karl Peters ('Empirical Theology in the Light of Science', *Zygon*, 1992, vol 27, 297-325).

301 See also the outlines of the debate in Barbour 1990:3-30 and Southgate 1999:3-16. In Barbour's oft-quoted terminology the outer columns are characterized by 'conflict', the second and fourth by 'independence', and the central column embraces his categories of 'dialogue' and 'integration'.

302 For a comprehensive list of scientists, theologians and organizations actively involved in the discourse see *Who's Who in Theology and Science* (1996, New York: Continuum).

303 Polkinghorne 1996:64-80

304 Southgate 1999:51-52. Also pages, 52-71, and Ward 1986:58-66.

305 Macquarrie 1971:308-310

306 Bowker 1987:61

307 Ward 1986:58-62, especially 61-62

308 Southgate 1999:52

309 Mooney 1991:310 (ref: note 12)

310 See Kees Niekerk's essay on critical realism in Gregersen & van Huyssteen 1998. The subject is discussed also in Polkinghorne, 1998:101-124; Peacocke 1993:11-19; and Barbour 1990:41-51.

311 Bowker 1995:155

312 They see this as a tightening grasp of reality that depends not on *picturability* but on *intelligibility*. They cannot actually picture an electron or photon yet they know a great deal about them.

313 The theme of S Hooke's *Alpha & Omega: The Pattern of Revelation* (1961).

314 Polkinghorne 1998:30-32

315 See his chapter in Gregersen & van Huyssteen 1998 (pp 181-231).

316 ibid, 1998:227

317 This model is linked to the *postfoundationalism* espoused by Wentzel van Huyssteen (in Gregersen & van Huyssteen 1998, and in his *Duet or Duel?*) which stands against the foundationalist theology of earlier times and the reductionist viewpoint of some contemporary science. It encourages a broader rationality that helps to promote dialogue between theology and science. It is sympathetic to both postmodern thought and critical realism where these affirm the context-relatedness and empirical aspects of knowledge. Also, his book *The Shaping of Rationality: Toward*

Interdisciplinarity in Theology & Science (1999) should be noted here as a major treatise on how argument, rhetoric and dialogue function in theology and in the sciences.

318 Polkinghorne 1998:81

319 Torrance 1985:79

320 ibid, pp 81, 83

321 Polanyi's epistemological work has been beautifully interpreted by Drusilla Scott in her *Everyman Revived: The Commonsense of Michael Polanyi*, and comprehensively treated in Joan Crewdson's *Christian Doctrine in the light of Michael Polanyi's Theory of Personal Knowledge*.

322 Crewdson 1994:3

323 Scott 1985:68

324 Crewdson 1994:58

325 The two criteria advocated by A J Ayer and Karl Popper respectively.

326 Macquarrie 1977:16

327 See Philip Clayton's paper 'The Inference to the Best Explanation' in *Zygon*, vol 32, (1997), no 3, pp 377-391.

328 Note the debate by C Isham and J Polkinghorne in Russell, *Quantum Cosmology and the Laws of Nature* (1993) pp 139-147.

329 Polkinghorne 1998:69. Divine temporality is discussed in detail in Polkinghorne 1989:77-84 and Peacocke 1993:128-133.

330 Peacocke 1993:121-123, 128

331 For a contrasting viewpoint (that God does indeed foreknow) see John Davis' paper 'Quantum Indeterminacy and the Omniscience of God' in *Science & Christian Belief*, vol 9, no 2 (Oct 1997), pp 129-144.

332 Described in David Pailin's *God and the Processes of Reality* (1989).

333 The emphasis is on *temporal*. God is thought of as constantly and uncoercively luring the world and its beings towards goodness and fulfillment.

334 Polkinghorne 1989:80

335 Peacocke 1993:128. As he points out, God's relation to time has for centuries entered into philosophical and theological discussions of major issues such as free will, predestination, divine impassibility, and the relation of time to eternity.

336 Barbour 1966:128-131

337 ibid, p 129
338 Barbour 1990:222-223
339 Discussed in Barbour 1966:439-463
340 Polkinghorne 1998:56
341 Polkinghorne 1996:31; Peacocke 1993:149
342 ibid, pp 33-34
343 Nancey Murphy in Russell, *Chaos and Complexity*, pp 338-343. Also see the related papers by Thomas Tracy and George Ellis in the same volume.
344 Polkinghorne 1998:60, 1996:31-32
345 Polkinghorne 1996:35. The idea of the universe as God's body has been canvassed to some extent, but it rather contradicts the traditional notion of separate existence as a gift from the One who has created the world *ex nihilo.*
346 Peacocke 1993:157-160 and in Russell, *Chaos & Complexity*, pp 282-283.
347 Polkinghorne 1996:36. He also refers to recent work of Ilya Prigogine who, in broadening the scope of certain equations describing the development in time of dynamical systems, shows that these then lead to a holistic account that is no longer rigidly deterministic. As in the case of quantum indeterminacy they indicate probabilities, not certitudes. See Polkinghorne 1998:63-67, especially pp 65-66.
348 Bearing in mind the notion of perichoresis (mutual indwelling) within the Trinity, note too the biblical references to the ongoing work of both the Father and the Son in the world (John 5:17 , and Revelation 21:5).
349 Taylor 1972:27-28
350 Peacocke 1993:174-175
351 Polkinghorne 1991:84
352 ibid, 1994:47
353 The word *kenosis* suggests a divine self-limiting that makes room for human free-will and nature's 'freedom of process'.
354 Moltmann 1995, 'Reflections on Chaos & God's Interaction with the World from a Trinitarian Perspective' in Russell, *Chaos & Complexity*, pp 205-210.
355 The Kabbala is a body of mystic oral tradition. Note also Peacocke's discussion of the self-limiting and self-giving of God, 1993:121-124.

356 Moltmann 1981:61

357 Ellis, *Before the Beginning* (1993) and 'The Theology of the Anthropic Principle' in Russell, *Quantum Cosmology and the Laws of Nature*, (1993) pp 363-399. See also *On the Moral Nature of the Universe* by Nancey Murphy & George Ellis (1996).

358 Ellis 1994, 'God and the Universe: Kenosis as the Foundation of Being', in *CTNS Bulletin* (Centre for Theology & the Natural Sciences), vol 14 no 2, pp 1-14.

359 Temple 1968:xxxi. Richard Harries makes the same point when he writes: 'What stops the God of Israel being experienced as sheer terror is his being as love and beauty. He is not just power but glory'. (Harries 1993:55).

360 In their experiments, scientists aim to set up *idealized* conditions in which perturbing influences are avoided as far as possible, the number of variables is minimized, and everything is kept as simple as possible.

361 George Ellis in Russell, *Quantum Cosmology and the Laws of Nature* (1993) pp 396-398

362 The ideas in this section are drawn largely from the author's paper 'Beauty in Physics and Theology' in *Journal of Theology for Southern Africa*, No. 94 (1996), pp 65-78.

363 Vanstone 1977:57-74

364 Weinberg 1993:117

365 Revelation 21:5

366 Romans 8:21. See the chapter on Eschatology in Polkinghorne (1994).

367 The need to find an appropriate balance is a keynote of Vincent Donovan's classic account of his experience as a missionary amongst the Masai, *Christianity Rediscovered* (1982).

368 *Creation* in the sense of *creatio continua*, the ongoing work of God throughout cosmic history.

369 Quoted in Polkinghorne 1991:11-12.

370 See for example Gunton 1991:14 (note 1) for a list of recent treatises from some of the main traditions.

371 Mooney 1991:319 (see note 12).

372 ibid, p 328

373 Polkinghorne 1991:1. He adds that theology must employ 'liturgy-assisted logic'.

374 Polkinghorne 1998:84-85. 'Immanent' refers to the divine nature as it is in itself, ontologically, whereas 'economic' refers to God's self-manifestation to the creation. It is the *economic* Trinity to which the three axioms of our Trinitarian cosmology refer.

375 'There is more to life, and the meaning of life, than science. There is also the world of myth. The word *myth* in the 19th century had not become synonymous with *false*, as it has for us. The concept and evaluation of myth has in fact had a very long history, during which its virtues and limitations have been threshed out ... above all in the 19th century when myth, positively understood, offered the great opportunity to claim that truth can be told as much through fiction as through scientific fact, ... as much through a poem as through a mathematical proof.' John Bowker, in a lecture at Gresham College, London, in February 1996.

376 Barbour 1990:206

377 Discussed in Stannard 1982:23-31.

378 Polkinghorne 1998:88-89

379 Here is a direct clash with the Pauline assertion that our bodies are going to die *because of sin* (Romans 8:10).

380 That which is last or final. Eschatology deals with ideas about 'the last things', such as death, judgement, and ultimate destiny.

381 Macquarrie 1977:78

382 Rolston 1987:286-293. Also see his paper, 'Does Nature need to be redeemed?' in *Zygon*, vol 29, no 2, pp 205-229.

383 Weinberg 1983:148-149. Recent astronomical evidence suggests the *un*liklihood of a 'Big Crunch' with its 'intolerable heat'.

384 Peacocke 1993:343. Note also pp 343-346.

385 Barbour 1990:152

386 ibid, p 241

387 Discussed in Polkinghorne 1994:162-175

388 ibid, 1996:56, 1994:170

389 ibid, 1994:163

390 Harries 1993:36. The Greek version of the Hebrew scriptures (used by most of the early Fathers) uses the word *kala* which embraces both meanings, *good* and *beautiful*.

391 Polkinghorne 1994:168. Patrick Sherry (1992:169) shares Polkinghorne's eschatological thinking to a large extent but gives the warning that the biblical warrant is slight. There is surely all

the more reason then for the further development of theologically based metaphysics.

392 Polkinghorne 1994:164. Note Romans 8:18-25.

393 ibid, p 167

394 Ward 1998:215-216

395 The ideas in this section and the next are drawn largely from the author's paper, 'Beauty in Physics and Theology' in *Journal of Theology for Southern Africa*, No. 94 (1996), pp 65-78.

396 Sherry 1992:21

397 Newbigin 1984:1

398 ibid, pp 21-27. He quotes Polanyi: 'The past four or five centuries, which have gradually destroyed or overshadowed the whole mediaeval cosmos, have enriched us mentally and morally to an extent unrivalled by any period of similar duration. But its incandescence has fed on the combustion of the Christian heritage in the oxygen of Greek rationalism, and when the fuel was exhausted, the critical framework itself burnt away.'

399 Newbigin's main books of this period are: *The Other Side of 1984* (1984); *Foolishness to the Greeks: The Gospel and Western Culture* (1986); and *The Gospel in a Pluralist Society* (1989). See also the author's paper 'The Gospel and Western Culture: on the Ideas of Lesslie Newbigin' in *Missionalia*, vol 27, no 1 (April 1999), pp 62-72.

400 Gunton 1993:102, 105

401 ibid, pp 50, 112-119

402 A term used in theology to describe the mutual indwelling, reciprocity and interanimation within the Trinity.

403 ibid, pp 144-5, 153

404 ibid, pp 150-151

405 ibid, p 177

406 ibid, p 191

407 Sherry 1992:5

408 ibid, p 2

409 Quoted in Sherry 1992:14-15. Note that where beauty has been treated theologically, it is the Son who has been the focus usually, rather than the Spirit. See especially the magisterial work of Hans Urs von Balthasar, *The Glory of the Lord: a Theological Aesthetics* (1961,1982) and Francesca Murphy's *Christ the Form of Beauty* (1995).

410 A phrase from W B Yeats' poem 'Easter 1916' (Harries 1993:47).

411 Harries writes of beauty as 'beckoning us to itself and pointing beyond itself to that which seems tantalizingly unattainable. It draws us to itself and through itself. ... If God is the giver of all good gifts and contains within himself all possible perfections, then he must be beauty as much as he is goodness and truth'. Harries 1993:42, 48. And on page 17 he explains: 'If I did not believe that God is the source and standard of all that I experience as beautiful, that he/she is beauty as much as truth and goodness, I would not be a religious believer at all'.

412 Chandrasekhar 1987:69

413 1 Peter 3:15

414 Bethge 1975:150, 153

415 Terrence Deacon (anthropologist) in July-August 1999 issue of *Science & Spirit*, pp 16-17.

416 For example, Wesley Granberg-Michaelson's *Tending the Garden: Essays on the Gospel and the Earth* (1987), Colin Russell's *The Earth, Humanity and God* (1994), and Ernst Conradie's *Hope for the Earth – Vistas on a New Century* (2000).

417 Hodgson and King 1985:1. Such openness is characteristic of the WCC leadership (Konrad Raiser in *International Bulletin of Missionary Research*, vol 18, April 1994) and is a keynote in the official position of the Catholic Church vis-a-vis other religions. It is superbly reflected, for example, in Decree Five of the Order of the Society of Jesus, *Our Mission and Interreligious Dialogue*, and in the new standard missiological work by Jacques Dupuis SJ, *Toward a Christian Theology of Religious Pluralism* (NY: Orbis 1997).

418 Christianity's stance vis-a-vis the ultimate destiny of humankind is usually characterized as variously *exclusive*, or *inclusive*, or *pluralist*. However, John Bowker names a fourth position, that of *differentialism*, which allows that differences of belief are real, that they imply different destinies, and that all of these may indeed be carried through to their different conclusions (Bowker 1995:181-183). In other words, perhaps the Lord God is pleased to affirm the particular destiny of each and seek to bring it to its own appropriate perfection.

419 Sarup 1993:153

420 A theme of Vincent Donovan's book, *Christianity Rediscovered: An Epistle from the Masai* (1982).

421 Here we should note the contrast between the Western world and Russia, a country which has not experienced this particular conflict. According to a Russian scholar at the Insitute of Philosophy of the Russian Academy of Science, Vladimir Katasonov, the fundamental reason is to be found in the specific understanding of the concept of *reason* in Eastern Christianity. Historically, reason was not seen as being opposed to faith; rather it was regarded as one of the faculties that grew out of the general root of faith. (From his paper 'Science and Religion in Russian Culture' in the magazine *Science & Spirit*, January/February 2000, p 16).

BIBLIOGRAPHY

Barbour I, *Issues in Science & Religion* (London: SCM Press, 1966)

Barbour I, *Religion in an Age of Science* (London: SCM Press, 1990)

Barrow J & Tipler F, *The Anthropic Cosmological Principle* (Oxford: Clarendon , 1986)

Bernal J D, *Science in History*, vol 2, 3rd ed (Penguin Books, 1965)

Berry R J, *God & Evolution* (London: Hodder & Stoughton, 1988)

Bethge E, *Bonhoeffer: Exile and Martyr* (London: Collins, 1975)

Blackmore V & Page A, *Evolution The Great Debate* (Oxford: Lion Publishing, 1989)

Bowker J W, *Licensed Insanities* (London: Darton, Longman & Todd, 1987)

Bowker J W, *Is God a Virus? Genes, Culture & Religion* (London: SPCK 1995)

Bowler P J, *Evolution: The History of an Idea* (Berkeley: U California Press, 1989, 2nd ed)

Brooke J H, *Science and Religion* (Cambridge U Press, 1991)
Browne J, *Charles Darwin: Voyaging* (London: Pimlico, 1995)
Butterfield H, *The Origins of Modern Science: 1300-1800* (London: Bell & Sons, 1957)
Capra F, *The Web of Life* (London: Flamingo, 1997)
Cohen H F, *The Scientific Revolution: a historiographical inquiry* (U Chicago Press, 1994)
Chandrasekhar S, *Truth and Beauty: Aesthetic Motivations in Science* (U Chicago Press, 1987)
Chown M, *Afterglow of Creation* (London: Random House, 1993)
Conradie E, *Hope for the Earth – Vistas on a New Century* (Cape Town: U Western Cape Publications, 2000)
Crewdson J, *Christian Doctrine in the Light of Michael Polanyi's Theory of Personal Knowledge* (Lampeter, Wales: Edwin Mellen Press, 1994)
Darwin C, *The Origin of Species* (Oxford U Press, 1996, ed Gillian Beer)
Davies P, *The Accidental Universe* (Cambridge U Press, 1982)
Davies P, ed, *The New Physics* (Cambridge U Press, 1989)
Davies P, *The Cosmic Blueprint* (Harmondsworth: Penguin, 1989)
Davies P, *The Mind of God* (Harmondsworth: Penguin, 1993)
Dawkins R, *The Selfish Gene* (London: Paladin/Granada Publishing, 1978)
Denton M, *Nature's Destiny* (New York: The Free Press, 1998)
Desmond A & Moore J, *Darwin* (Harmondsworth: Penguin, 1992)
Donovan V, *Christianity Rediscovered: An Epistle from the Masai* (London: SCM, 1982)
Ellis G, *Before the Beginning* (Boyars/Bowerdean 1993)
Fiddes P, *The Creative Suffering of God* (Oxford: Clarendon Press, 1988)
Gillispie C C, *Genesis and Geology* (New York: Harper Torchbooks, 1959)
Gleick J, *Chaos: Making a New Science* (London: Penguin, 1987)
Gonzalez J L, *The Story of Christianity* (San Francisco: Harper Collins, 1985)
Goodman D C & Russell C A, *The Rise of Scientific Europe* (Hodder & Stoughton, 1991)
Goodman D C (ed), *Science & Religious Belief 1600-1900: A Selection of Primary Sources* (Open U Press, 1973)
Granberg-Michaelson W (ed), *Tending the Garden: Essays on the Gospel and the Earth* (Grand Rapids: Eerdmans, 1987)
Greene B, *The Elegant Universe* (London: Jonathan Cape, 1999)
Greene M T, *Geology in the Nineteenth Century* (Ithaca: Cornell U Press, 1982)

Gregersen N H & van Huyssteen J W, *Rethinking Theology and Science* (Grand Rapids: Eerdmans, 1998)
Gunton C, *Enlightenment and Alienation* (London: Marshall Morgan & Scott 1985)
Gunton C, *The Promise of Trinitarian Theology* (Edinburgh: T & T Clark, 1991)
Gunton C, *The One, the Three and the Many* (Cambridge U Press, 1993)
Harries R, *Art and the Beauty of God* (London/New York: Mowbray, 1993)
Hawking S, *A Brief History of Time* (London: Bantam, 1988)
Hodgson P C & King R H, *Christian Theology: An Introduction* (Minneapolis: Fortress Press, 1985, 2nd ed)
Hooykaas R, *Religion and the Rise of Modern Science* (Edinburgh: Scottish Academic, 1972)
Hopper J, *Understanding Modern Theology: Cultural Revols. & New Worlds* (Minneapolis: Fortress, 1987)
Howard J, *Darwin* (Oxford U Press, 1982)
Kauffman S, *At Home in the Universe* (NY: Oxford U Press, 1995)
Lash N, *The Beginning and the End of 'Religion'* (Cambridge U Press, 1996)
Lindberg D C & Numbers R L (eds), *God & Nature* (U California Press, 1986)
Macquarrie J, *Twentieth-Century Religious Thought* (London: SCM Press, 2nd ed, 1971)
Macquarrie J, *Principles of Christian Theology* (London: SCM, 2nd ed, 1977)
Moltmann J, *The Crucified God* (London: SCM Press, 1974)
Moltmann J, *The Trinity & the Kingdom of God* (London: SCM Press, 1981)
Monod J, *Chance and Necessity* (NY: Vintage Books, 1972)
Moore J, *The Post-Darwinian Controversies* (Cambridge U Press, 1979)
Murphy F A, *Christ the Form of Beauty* (Edinburgh: T&T Clark, 1995)
Murphy N & Ellis G, *On the Moral Nature of the Universe* (Minneapolis: Fortress, 1996)
Nebelsick H, *Theology and Science in Mutual Modification* (New York: Oxford U Press, 1981)
Neill S, *Anglicanism* (London: Penguin Books, 1958)
Newbigin L, *The Other Side of 1984* (WCC, Geneva, 1984)
Newbigin L, *Foolishness to the Greeks: The Gospel & Western Culture* (London: SPCK, 1986)
Newbigin L, *The Gospel in a Pluralist Society* (Eerdmans and WCC, 1989)

Open U Course AMST 283, *Science & Belief: from Copernicus to Darwin* (Open U Press, 1974):
Hooykaas R, 'The Impact of the Copernican Transformation', Unit 2
Goodman DC, 'Galileo and the Church', Unit 3
Goodman DC, 'God & Nature in the Philosophy of Descartes' Unit 4
Brooke JH, 'Newton and the Mechanistic Universe', Unit 5
Pailin D, *God and the Processes of Reality* (London: Routledge, 1989)
Peacocke A, *Creation and the World of Science* (Oxford U Press, 1979)
Peacocke A, ed, *The Sciences and Theology in the 20th Century* (London: Routledge & Kegan Paul, 1981)
Peacocke A, *Theology for a Scientific Age* (London: SCM, 1993)
Polanyi M, *Personal Knowledge* (New York/London: Routledge, 1958)
Polkinghorne J, *The Way the World is*, (London: SPCK, 1983)
Polkinghorne J, *One World* (London: SPCK, 1986)
Polkinghorne J, *Science and Creation* (London: SPCK, 1988)
Polkinghorne J, *Science and Providence* (London: SPCK, 1989)
Polkinghorne J, *Reason and Reality* (London: SPCK, 1991)
Polkinghorne J, *Science and Christian Belief* (London: SPCK, 1994); also published as *The Faith of a Physicist* (Princeton U Press, 1994)
Polkinghorne J, *Scientists as Theologians* (London: SPCK, 1996)
Polkinghorne J, *Belief in God in an Age of Science* (New Haven:Yale U Press, 1998)
Poole M, *Beliefs and Values in Science Education* (Open U Press, 1995)
Prigogine I & Stengers I, *Order out of Chaos* (NY: Bantam Books, 1984)
Richardson W M & Wildman W J, eds, *Religion and Science* (New York/London: Routledge, 1996)
Ridley M, ed, *Evolution* (Oxford U Press, 1997)
Rolston H, *Science and Religion* (NY: Random House, 1987)
Rolston H, *Conserving Natural Value* (New York: Columbia U Press, 1994)
Rowse A L, *The Elizabethan Renaissance*, vol 3 (London: Macmillan, 1972)
Russell C A (ed), *Science and Religious Belief: recent historical studies* (Open U Press, 1973)
Russell C A, *Science & Social Change: 1700-1900* (London: Macmillan, 1983)
Russell C A, *Cross-currents: interactions between Science & Faith* (London: Inter-Varsity Press, 1985, and Christian Impact Press, 1995)
Russell C A, *The Earth, Humanity and God* (London: University College London Press, 1994)

Russell R J et al, eds, *Physics, Philosophy & Theology* (Vatican Observ, 1988)
Russell R J et al, eds, *Quantum Cosmology & the Laws of Nature* (Vatican Observatory/CTNS, 1996)
Russell R J et al, eds, *Chaos & Complexity* (Vatican Observatory/CTNS,1995)
Sarup M, *Introd Guide to Post-Structuralism & Postmodernism* (Harvester Wheatsheaf, 1993)
Scott D, *Everyman Revived: The Common Sense of Michael Polanyi* (Book Guild Ltd, 1985)
Shapin S, *The Scientific Revolution* (U Chicago Press, 1996)
Sherry P, *Spirit and Beauty: An Introduction to Theological Aesthetics* (Oxford: Clarendon Press, 1992)
Sorell T, *Descartes* (Oxford U Press, 1987)
Southgate C, ed, *God, Humanity & the Cosmos* (Edinburgh: T&T Clark, 1999)
Stannard R, *Science and the Renewal of Belief* (London: SCM Press, 1982)
Stewart I, *Life's Other Secret: the New Mathematics of the Living World* (London: Penguin, 1999)
Tarnas R, *The Passion of the Western Mind* (Random House, 1991)
Taylor J V, *The Go-Between God* (London: SCM Press, 1972)
Temple W, *Readings in St John's Gospel* (London: MacMillan, 1968)
Torrance T, *Reality and Scientific Theology* (Edinburgh: Scottish Academic, 1985)
Van Huyssteen J W, *Duet or Duel* (London: SCM Press, 1998)
Van Huyssteen J W, *The Shaping of Rationality: Toward Interdisciplinarity in Theology & Science* (Grand Rapids: Eerdmans, 1999)
Vanstone W H, *Love's Endeavour, Love's Expense* (London: Darton, Longman and Todd, 1977)
Vidler A R, *The Church in an Age of Revolution* (London: Penguin, 1961)
Von Balthasar H U, *The Glory of the Lord: A Theol. Aesthetics* (Edinburgh: T&T Clark, 1982)
Von Grunebaum G E, *Islam: Essays in the Nature & Growth of a Cultural Tradition* (London: Routledge, 1969)
Ward K, *The Turn of the Tide* (London: BBC, 1986)
Ward K, *God, Faith & the New Millenium* (Oxford: Oneworld, 1998)
Weinberg S, *The First Three Minutes* (London: Fontana, 1983)
Weinberg S, *Dreams of a Final Theory* (London: Vintage, 1993)

Wertheim M, *Pythagoras' Trousers: God, Physics & the Gender Wars* (Times Books, 1995)
Westfall R S, *Never at rest* (Cambridge U Press, 1980)
White M, *Isaac Newton: the Last Sorcerer* (Fourth Estate, 1997)
Whitehead A N, *Science and the Modern World* (Cambridge UP, 1927, and NY:Mentor Books, 1948)

INDEX

T

U-V

W-Z

Lightning Source UK Ltd.
Milton Keynes UK
UKOW04f1851070216

267905UK00001B/98/P